Ion Beams — New Applications
from Mesoscale to Nanoscale

MATERIALS RESEARCH SOCIETY
SYMPOSIUM PROCEEDINGS VOLUME 1354

Ion Beams — New Applications from Mesoscale to Nanoscale

Symposium held April 25–29, 2011, San Francisco, California, U.S.A.

EDITORS

John Baglin

IBM Almaden Research Center
San Jose, California, U.S.A.

Daryush ILA

Fayetteville State University
Fayetteville, North Carolina, U.S.A.

Giovanni Marletta

Universita degli Studi di Catania
Catania, Italy

Ahmet Öztarhan

Ege University
Izmir, Turkey

Materials Research Society
Warrendale, Pennsylvania

CAMBRIDGE
UNIVERSITY PRESS

CAMBRIDGE
UNIVERSITY PRESS

Shaftesbury Road, Cambridge CB2 8EA, United Kingdom

One Liberty Plaza, 20th Floor, New York, NY 10006, USA

477 Williamstown Road, Port Melbourne, VIC 3207, Australia

314–321, 3rd Floor, Plot 3, Splendor Forum, Jasola District Centre, New Delhi – 110025, India

103 Penang Road, #05–06/07, Visioncrest Commercial, Singapore 238467

Cambridge University Press is part of Cambridge University Press & Assessment, a department of the University of Cambridge.

We share the University's mission to contribute to society through the pursuit of education, learning and research at the highest international levels of excellence.

www.cambridge.org
Information on this title: www.cambridge.org/9781605113319

First published 2012

A catalogue record for this publication is available from the British Library

ISBN   978-1-605-11331-9   Hardback

CONTENTS

*Invited Paper

FORUM REPORT

PREFACE

Symposium II, "Ion Beams — New Applications from Mesoscale to Nanoscale" was held at the 2011 MRS Spring Meeting, April 25-29, 2011 in San Francisco, California.

This symposium welcomed presentations on ion-beam engineering and characterization of materials properties, structure, topography, or functionality, spanning dimensions from the mesoscale to the nanoscale. Indeed, while the unique capabilities of ion-beam techniques in the diverse emerging fields of nanoscience and nanotechnology are fast becoming critical for many new applications, the flexibility of ion-beam techniques now enables the development of new tools that can integrate tailoring of nanoscale patterns and structures with unique in-situ imaging and analysis – as indicated by the wealth of research reports presented in this volume. The recent evolution of such instrumentation has energized new programs, both basic and applied, in fast-developing areas ranging over advanced semiconductor integration, information storage, sensors, plasmonics, molecular engineering, biomaterials, and many aspects of the development of alternative energy resources.

In a field displaying such rapid evolution on many fronts, it is appropriate for us to pause occasionally and review the overall state of the field, and its emerging opportunities and challenges. Two special Forum sessions, late in the Symposium, (titled "Ion Beam Institute"), were held to facilitate such consideration. They yielded lively discussion and remarkable consensus of opinion about the direction and priorities of the field for the future. These sessions are summarized in detail in the concluding paper of this volume. That report displays a dynamic state of the field, especially with respect to interdisciplinary applications, and notably for the bio sciences, where ion beam techniques have much to offer in new emerging basic research and applications.

The organizers would like to take this opportunity to thank the sponsors of this Symposium, (NASA and National Electrostatics Corporation) for their generous financial support.

John Baglin
Daryush ILA
Giovanni Marletta
Ahmet Öztarhan

September 2011

MATERIALS RESEARCH SOCIETY SYMPOSIUM PROCEEDINGS

MATERIALS RESEARCH SOCIETY SYMPOSIUM PROCEEDINGS

Prior Materials Research Society Symposium Proceedings available by contacting Materials Research Society

Reviews and Research Reports

Mater. Res. Soc. Symp. Proc. Vol. 1354 © 2011 Materials Research Society
DOI: 10.1557/opl.2011.1382

Universal Biomolecule Binding Interlayers Created by Energetic Ion Bombardment

Prof Marcela M.M. Bilek[1], Prof David R. McKenzie[1], Dr Daniel V. Bax[1,2], Dr Alexey Kondyurin[1], Dr Yongbai Yin[1], Dr Neil. J. Nosworthy[1,3], Ms Stacey Hirsh[1], Dr Keith Fisher[4], Prof Anthony S. Weiss[2]

[1] School of Physics, University of Sydney, NSW 2006 Australia
[2] School of Molecular Biosciences, University of Sydney NSW 2006 Australia
[3] School of Medical Sciences, University of Sydney, NSW 2006 Australia
[4] School of Chemistry, University of Sydney, NSW 2006 Australia

ABSTRACT

The ability to strongly attach biomolecules such as enzymes and antibodies to surfaces underpins a host of technologies that are rapidly growing in utility and importance. Such technologies include biosensors for medical and environmental applications and protein or antibody diagnostic arrays for early disease detection. Emerging new applications include continuous flow reactors for enzymatic chemical, textile or biofuels processing and implantable biomaterials that interact with their host via an interfacial layer of active biomolecules. In many of these applications it is desirable to maintain physical properties of an underlying material whilst engineering a surface suitable for attachment of proteins or peptide constructs. Nanoscale polymeric interlayers are attractive for this purpose.

We have developed interlayers[1] that form the basis of a new biomolecule binding technology with significant advantages over other currently available methods. The interlayers, created by the ion implantation of polymer like surfaces, achieve covalent immobilization on immersion of the surface in protein solution. The interlayers can be created on any underlying material and ion stitched into its surface. The covalent immobilization of biomolecules from solution is achieved through the action of highly reactive free radicals in the interlayer.

In this paper, we present characterisation of the structure and properties of the interlayers and describe a detailed kinetic model for the covalent attachment of protein molecules directly from solution.

INTRODUCTION

Environmental biosensors[2], antibody microarrays for early and precise disease diagnosis [3]and bio-mimetic surface coatings for medical implants[4, 5] require functional[4, 6] immobilised proteins, such as enzymes and antibodies. The capability to robustly immobilise all proteins expressed in a cell would enable "reverse phase" microarrays [7]. The biological responses to implanted biomedical devices could potentially be controlled with a conformal coverage of bioactive proteins or peptide segments [4, 8-10]. Simple techniques to achieve strong and functional binding[9] that perform well across a wide range of proteins and on a wide range of surfaces are required to facilitate the development of medical and sensing technologies that call for immobilised biomolecules with high functionality.

The *conformation* or shape of a biomolecule is the configuration in space that minimises the energy of the protein and its immediate environment. The *native conformation* is typically required to enable the protein molecule's biological function. When attempting to immobilise functional protein molecules we must therefore consider the effects of a proximate surface on the molecule's conformation. Figure 1 depicts schematically a protein molecule as a chain of covalently linked amino acids (a) that adopts configuration (b) to minimise free energy in solution. It results in predominantly hydrophilic amino acid residues (shown in white) on the surface of the molecule exposed to solution and hydrophobic residues (shown in black) in its centre. This energy minimising structure is known as the protein molecule's *native conformation*. In the presence of a hydrophobic surface there are other configurations that allow for energy minimisation of the system. These are conformations (c) in which the protein exposes predominantly hydrophobic residues to the surface while exposing predominantly hydrophilic residues to the surrounding water molecules. In the presence of a hydrophilic surface there are no significant energetic advantages to be gained from conformational changes but there are also no forces for adhesion. Strongly hydrophilic surfaces therefore to not adhere protein molecules from solution. An ideal immobilisation surface should be hydrophilic to avoid destruction of protein native conformations and must therefore rely on forces other than physical interactions to immobilise protein molecules. Covalent bonding has been demonstrated as a way to achieve strong immobilisation. In previously reported covalent coupling approaches, specially synthesised linker molecules (d) that have reactive end groups to form bonds with specific groups on amino acid side chains of the protein molecule and with reactive sites on the surface

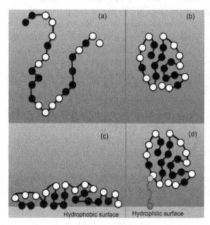

Figure 1: A protein, made up of a sequence of amino acids (a), covalently linked by amide bonds, adopts a configuration in solution (b), that minimises the Gibbs free energy of the system including the surrounding solvent. In the presence of hydrophobic surface (c), other configurations will represent energy minima. These configurations place hydrophobic residues on the surface and expose hydrophilic residues to solution. The proximity of a hydrophilic surface (d), does not give rise to any new configurations that would lead to energy minimisation of the system.

are required. Although highly effective, this method of immobilisation is rarely implemented in non-research applications because it involves complex and time consuming wet chemical steps and limits the surfaces that can be used for immobilisation. Despite the fact that physical approaches yield unstable, non-robust and variable attachment that compromises the native conformation of immobilised proteins they dominate in commercially realised applications due to their simplicity and low cost.

Free radicals are often seen as the "bad guys" of biology, having been implicated in aging [11] and in many diseases arising from the malfunctioning of proteins[12]. However, evidence is emerging that shows their potential to do useful tasks such as immobilising proteins on surfaces while preserving protein function. An effective method of creating buried radicals is to treat an organic polymer with energetic ions[13, 14]. Treating PTFE with energetic ions resulted in a very reactive surface which formed strong bonds with applied adhesives[15]. Organic polymer surfaces treated by energetic ions, either post-formation or during their deposition, strongly attached protein molecules from solution[10, 16-29]. Occasional speculation about a link between strong binding and free radicals can be found in the literature[16, 30, 31] but until now there has been little direct evidence and the crucial role of radical migration was not recognised.

In this paper, we review a new approach to the surface immobilisation of functional biomolecules[32] that relies on a reservoir of mobile free radicals, created under the surface during the ion irradiation of polymeric materials.

MATERIALS AND METHODS

Polymers for plasma processing were purchased from Goodfellow, UK. Enzymes for surface immobilization were purchased from Sigma Aldrich and Human recombinant tropoelastin was expressed in house as described in [33].

Plasma immersion ion implantation of polymer samples was carried out in an inductively coupled radio-frequency plasma. Samples were mounted on a stainless steel holder, with a stainless steel mesh of 150 mm diameter, electrically connected to the holder and placed 45 mm in front of the sample surface. Acceleration of ions from the plasma was achieved by the application of -20 kV high voltage bias pulses of 20 ms duration to the sample holder at a frequency of 50 Hz, unless otherwise specified. The sample holder was earthed between the pulses. The samples were treated for durations of 20 - 1600 s (800 seconds unless otherwise specified), corresponding to implantation ion fluence of $0.02 - 2.0 \times 10^{16}$ ions/cm^{-2}.

For deposition of ion treated plasma polymers on non-polymeric surfaces acetylene (purity 98% and flow rate 10 sccm) was injected as the polymer precursor and mixed with argon and nitrogen unless otherwise specified. A radio frequency (RF) electrode (bottom electrode) was used to generate plasma while the substrate holder (top electrode) was biased by a dc pulsed voltage source. Unless stated otherwise, the pulse voltage was -200 V at 10 kHz with a duty cycle of 10%. Substrates used included 316L stainless steel foil (25 μm thick) for ELISA and thin (12 μm thick) polyimide film for ESR measurements.

Electron spin resonance (Bruker Elexsys E500 EPR spectrometer) was used to detect unpaired electrons. The spectrometer operated in X band with a microwave frequency of 9.33GHz and a centre field of 3330G, at room temperature and was calibrated using a weak pitch sample in KCl and also with DPPH (a,a'–diphenyl-b-picrylhydrazyl). Samples for analysis were cut to a size of 5cmx5cm, rolled and placed into a quartz tube of 5 mm diameter.

Wettability was measured using the sessile drop method (Kruss DS10) with de-ionised water, glycerol, methylene diiodine and formamide. The surface energy and its components (polar and dispersic parts) were calculated using the Owens-Wendt-Rabel-Kaelble model. Atomic force microscopy (AFM) images of samples were collected on a PicoSPM instrument in tapping mode and analysed using the WSxM software (version 3, Nanotec Electronica S.L.).

Attenuated total reflection Fourier transform infra-red (ATR-FTIR) spectroscopy was used to study changes in surface chemistry and detect the presence of immobilised protein. We used a Digilab FTS7000 FTIR spectrometer fitted with an ATR accessory (Harrick, USA) with trapezium Germanium crystal and incidence angle of 45°. The thickness of the measured layer was 400-800 nm. Samples undergoing SDS treatment prior to FTIR measurement were immersed for 1 hour in sodium dodecyl sulfate (SDS) 2% solution at 70C unless stated otherwise and then washed in milliQ-water 3 times (20 mins each) at 23C. All samples were dried overnight prior to measurement. The intensity of the 1462 cm^{-1} methylene group vibration was used as an internal standard for normalisation. Changes in C=O group concentration was determined from the absorption intensity at 1720 cm^{-1}. The Amide A, I and II absorptions from the protein backbone were used to quantify protein. Deconvolution of the amide I line using the GRAMS software provided information on changes in protein conformation with Gaussian peaks at 1,694, 1,684, and 1,674 cm^{-1} representing β-turns; 1,661 and 1,649 cm^{-1} for α-helices;1,635 cm^{-1} for random structures; and 1,623 cm^{-1} for β-sheets.

A tetramethylbenzidene (TMB) assay was used to assess the activity of horse radish peroxidase immobilised on the polymer surfaces. After overnight incubation in the HRP buffer solution, samples were washed six times for 20 min in fresh buffer solution. And then clamped between two stainless steel plates separated by an O ring (inner diameter 8mm) which sealed to the sample surface. The top plate contained a 5-mm-diameter hole, enabling the addition of 75 microlitres of TMB (3,3',5,5' tetramethylbenzidine liquid substrate system for ELISA—Sigma T0440), an HRP substrate containing 0.012% hydrogen peroxide. After 30 s, 25 microlitre aliquots were taken and added to 50 microlitres of 2M HCl in a 100-microlitre cuvette, another 25 microlitres of TMB was added to bring the volume to 100 microlitres. The optical density (OD) at a wavelength of 450 nm was measured in transmission using a Beckman DU530 Life Science UV/vis spectrophotometer.

ELISA was used to quantify surface attached tropoelastin with primary anti-body, mouse anti-elastin antibody (BA-4), and goat anti-mouse IgG-HRP conjugated secondary antibody. The samples undergoing SDS treatment prior to ELISA were transferred to 5% SDS (w/v) in PBS and incubated at 90°C for 10 min. Non-specific binding to the polymer was blocked with 3% (w/v) bovine serum albumin (BSA) in PBS. The HRP label was detected by adding 0.75ml ABTS solution (40mM ABTS [2,2'-azino-bis (3-ethylbenzthiazoline-6-sulphonic acid)] in 0.1M NaOAc, 0.05M NaH$_2$PO$_4$, pH5, containing 0.01% (v/v) H$_2$O$_2$) and measuring the absorbance at 405nm using a plate reader.

Results and Discussion

Free radicals have an unpaired electron and therefore an associated electron spin making them extremely reactive. Our strategy is to use plasma processes involving ion implantation to modify polymeric materials so that they contain a reservoir of stable and mobile free radicals. The stabilisation is achieved by the formation of π conjugated clusters on which unpaired electrons are delocalised and mobility through the structure occurs through hopping between adjacent clusters. This structure can be achieved in two ways. The first (shown schematically in Figure 2(a)) is by treating a carbon-based polymer using a technique known as plasma immersion ion implantation. Ions are accelerated into the surface by an electric field of the plasma sheath that forms around the sample holder. The second method (shown schematically in Figure 2(b)) is by the deposition of a plasma polymer film under moderate ion bombardment (up to 1 kV). This method can be used on non-polymeric materials. The electron spin density created by ion treatment is quantified using electron spin resonance (ESR). High concentrations of unpaired electrons are measured in both ion implanted (Figure 2(c)) and plasma deposited polymers (Figure 2(d)) for periods of many months after the ion treatment.

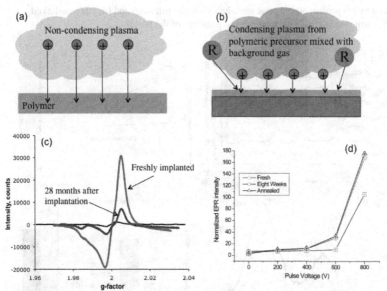

Figure 2: Schematic diagrams showing the methods we have employed to create free radical reservoirs. When the underlying material is a polymer (a) a high voltage (~20 kV) is applied in pulsed mode to accelerate ions from a surrounding non-condensing plasma. If the material is not a carbon rich polymer a plasma polymer film (b) is deposited under moderate ion bombardment applied in the form of a pulsed bias (up to 1 kV) during the deposition. (c) ESR signal from PIII treated PTFE showing the presence of free radicals. The red curve is for a freshly treated sample (20 mins after treatment), the blue curve for a 28-month old sample and the black signal is obtained from an untreated control sample. (d) ESR signal intensity as a function of pulse bias voltage applied during plasma polymer deposition. Measurements were taken 4 days after deposition (red circles), 8 weeks after deposition both before (blue squares) and after (black triangles) reactivation by annealing.

Figure 3(a) shows infra-red spectra revealing typical structural changes that occur in polymers upon ion implantation as a function of the ion implantation treatment time. The gradual reduction of the intensity of the characteristic polystyrene vibrations (1-5) shows that the polymer macromolecules are broken up and replaced by a carbonised structure that oxidises in the laboratory atmosphere (rise of peaks 8 and 9 respectively)[34, 35]. Figure 3(b) shows the evolution in treated polymers that typically occurs after removal from the plasma chamber. Reactions of free radicals, created under the surface by the ion impacts, with species in the atmosphere result in changes in surface energy and C=O IR adsorption bands that are correlated with changes in the electron spin density. The concentration of C=O groups on the surface increases during exposure to atmosphere because atmospheric oxygen reacts with radicals[36] in the polymer. The surface energy measured at the first time point is significantly higher than that of an untreated surface and then progressively decreases as high-energy radicals react with environmental oxygen. Such surfaces strongly attach protein molecules upon incubation in protein containing solution. There is minimal loss of the attached protein after washing with sodium dodecyl sulphate (SDS), a detergent capable of disrupting non-covalent interactions [16], as observed by (c) protein amide peak absorbances in the infra-red and (d) enzyme-linked immunosorbent assays (ELISA) detecting the presence of the protein.

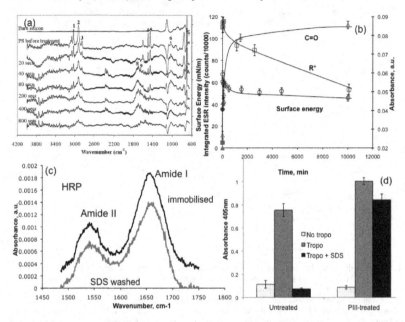

Figure 3: ATR-FTIR spectra of ion implanted polystyrene (a) show that the polystyrene macromolecular structure is gradually degraded, dehydrogenated and replaced by a carbonized structure that is oxidized on exposure to atmosphere. The spectra (top to bottom) are taken from a bare silicon wafer; untreated polystyrene; polystyrene treated for 20, 40, 80, 200, 400 and 800 seconds. (b) The surface energy, normalized carbonyl group absorbance from FTIR ATR and ESR intensity of free radicals as a function of storage time for PIII treated LDPE (20 keV, 20μs pulses at 50Hz in nitrogen). (c) FTIR ATR spectra of HRP enzyme immobilised on UHMWPE treated as in (b) to a fluence of 1e16 ions/cm² (UHMWPE background subtracted). Protein attached during incubation (black spectrum, upward shifted) is virtually 100% retained after SDS washing (red spectrum) [2% SDS solution, 70C for 1 hour]. (d) ELISA confirms that tropoelastin physisorbed from 20 μg/ml solution (grey bars) is retained only on the PIII treated PTFE (right) after SDS washing. The white bars indicate the assay background signal without tropoelastin in solution.

Figure 4 shows a characteristic curve describing the resistance to elution for a SDS washing protocol used by Kiaei et al [31] to remove albumin from a range of untreated polymers and plasma polymer surfaces. A clear trend (shown by the curve) with surface energy is apparent, with the strongest adsorption on the most hydrophobic (lowest energy) surfaces. Note that the room temperature SDS protocol employed by Kiaei et al does not remove all of the physisorbed protein. Data from plasma immersion ion implantation (PIII) treated polymers (red squares) and untreated polymers (blue diamonds) where a range of washing protocols were employed is also shown. Aggressive SDS protocols at 70-90C completely elute protein from very hydrophobic surfaces, such as PTFE. The PIII treated surfaces typically show 50-100% protein retention

8

despite being quite hydrophilic. This indicates that physical forces alone cannot be responsible for the robust protein attachment observed on the ion implanted surfaces implying that a covalent linkage is formed. The ability to covalently immobilise onto a strongly hydrophilic surface is a key advance that allows the retention of protein bioactivity by providing an environment conducive to retaining native conformations.

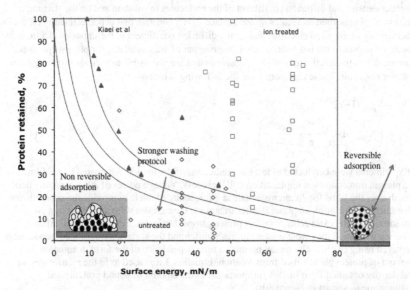

Figure 4: Percent of protein retained after SDS washing (various solution strengths and temperatures) as a function of surface energy for polymeric surfaces. Data is taken from Kaiei *et al* [31] (green triangles) and current work (red squares and blue diamonds). The top curve shows is a trend curve for the data of Kaiei *et al* and the lower curves show shifts in the trend expected for more rigorous SDS washing. Untreated controls washed with stronger SDS protocols are shown as blue diamonds. The points (red squares and one green triangle) lying above and to the right of the trend curves show exceptional protein retention given the hydrophilic nature of these surfaces.

We propose a model in which the covalent binding takes place via a reaction between an amino acid residue on the protein and a free radical on the ion treated polymer surface that is created by the diffusion to the surface of an unpaired electron from a reservoir below the surface. The first step is the physisorption of a protein on the surface and the second step is the formation of a covalent bond between a protein residue and a radical group (illustrated schematically in Figure 5(a)). There are two relevant time constants, one for the diffusion of proteins in solution to the surface and the second for the diffusion of the unpaired electrons from the reservoir to the surface. These processes are governed by the following two coupled differential equations:

$$\frac{dNp}{dt} = \frac{(N_{psites} - Np)}{\tau_1} \tag{1}$$

9

$$\frac{dNc}{dt} = \frac{(FNp - Nc)}{\tau_2} \tag{2}$$

where N_p is the number of physisorbed protein molecules per unit area and N_c is the number of covalently immobilised protein molecules per unit area. t_1 is a constant that depends linearly on the number density and diffusion coefficient of the molecules in solution and on the sticking coefficient of physisorbed molecules on the surface. t_2 is a constant that depends linearly on the number density of unpaired electrons and on the diffusion coefficient of the unpaired electrons in the modified region of the polymer . N_{psites} is the number of sites available for physisorption per unit area and F is the fraction of physisorption sites that are accessible to radicals diffusing from the interior reservoir. These equations have the following solutions:

$$N_p = N_{psites}(1 - e^{-t/\tau_1}) \tag{3}$$

$$N_C = FN_{psites}\left(1 - \frac{\tau_1 e^{-t/\tau_1}}{\tau_1 - \tau_2} - \frac{\tau_2 e^{-t/\tau_2}}{\tau_2 - \tau_1}\right) \tag{4}$$

Experiments were conducted to test the predictions of the model. We treated PTFE films with a plasma immersion ion implantation (PIII) process. Voltage pulses of 20kV were applied to a mesh over the films for 20 microseconds at a frequency of 50Hz to provide the energetic ion bombardment from a nitrogen plasma. After treatment, the samples were incubated in a 20 mg/ml solution of the extra cellular matrix protein, tropoelastin, for a range of times. After removal from solution they were washed in fresh buffer and the surface immobilized tropoelastin was assayed using ELISA. The model predicts a time dependence of the form in equation 3 of the amount of protein physisorbed from solution. Equation 3 was used to fit the relationship of optical density obtained from ELISA (proportional to the amount of bound protein) and incubation time as shown in Figure 5(b).

In parallel, a group of equivalent samples was subjected to rigorous SDS washing prior to ELISA detection of the tropoelastin. The protein detected in this case is covalently bound and thus would be expected to show the time dependence predicted by equation 4. A fit of this data by equation 4 is shown in Figure 5(b). The parameters determined in the fitting of equation 3 (N_{psites} and t_1) were used in the fit of equation 4 with F=1, leaving only one free parameter, the time constant for covalent binding, t_2. The time constants, t_1 and t_2 were determined to be 4.3+1.2 minutes and 35+9 minutes respectively. The shape of the experimental curve for covalent attachment is distinctly different from that for physisorption, especially in its behaviour at short incubation times. The time dependences of both the physisorption and covalent attachment are well reproduced by the theory, confirming the presence of processes for covalent attachment with two different time constants.

Modelling the motion of unpaired electrons in the reservoir by kinetic theory gives the following equation for the time constant of covalent binding of adsorbed molecules, t_2:

$$\tau_2 = \frac{4FN_{psites}}{n_r \bar{v}_r S_r} \tag{5}$$

where n_r is the number density of unpaired electrons, $\bar{v}_r$ the mean velocity associated with their diffusion, and S_r the probability of interaction with an adsorbed protein that forms a covalent bond. The number density of free radicals in the reservoir beneath the surface was observed to decay with time as shown in Figure 2(c & d). This leads to an increase in t_2 for an aged sample. The increase in t_2 is expected to be greater than predicted on the basis of the reduction in n_r alone since both $\bar{v}_r$ and S_r may decrease with time. S_r would decrease as some of the surface becomes passivated by adsorption of atmospheric contaminants during storage. Radicals arriving at the surface at a site covered by contaminants will covalently bind to these rather than to a protein molecule. $\bar{v}_r$ may decrease over time because the unpaired electrons in environments allowing the highest mobility will have the highest rate of quenching.

To test the dependence predicted by the model on the sample's age, we compare in Figure 5(c) the dependence of the amount of protein covalently attached on incubation time for new and aged (448 days) PIII treated PTFE films. The time constant, t_2, for free radical binding found by fitting equations 3 and 4 to the data for the old sample was 3 ± 1 days, two orders of magnitude greater than for the new sample. On the basis of the estimated reduction in free radical number density, n_r, obtained from the ESR spectra of Figure 2(c) there would be no more than a 10 fold increase in t_2. The remainder of the change is ascribed to decreases in $\bar{v}_r$ and S_r.

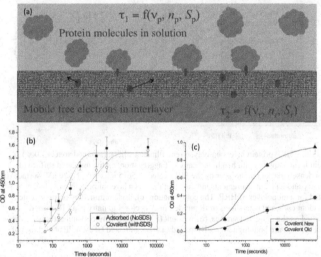

Figure 5: (a) A schematic diagram showing the model for the covalent attachment of protein molecules to the surface. There are two relevant time constants, one (τ_1) governing the adsorption of a protein layer on the surface and the second (τ_2) governing the flow of free electrons to the surface. Panel (b) shows the dependence of tropoelastin coverage on incubation time for PIII treated PTFE. The lines show the fits of equations 3 and 4 to the amounts adsorbed and covalently immobilised. Panel (c) shows how the time constant for covalent immobilisation increases with sample age.

Figure 6 shows the characteristics of covalently attached protein layers on polymer surfaces treated with energetic ions. A comparison with a protein layer physically adsorbed onto an

untreated polystyrene surface shows that a densely packed monolayer forms on the treated
surface (Figure 6 (b)) while a layer with significant void fraction is observed on untreated
polystyrene (Figure 6 (a)). The conformation of the protein on the treated surface is closer to a
native conformation than that for the untreated polystyrene surface (Figure 6(c)). HRP enzyme
bound to the treated surface shows significantly better resistance to removal by washing and
better retention of enzyme activity than that adsorbed onto untreated UHMWPE (Figure 6(d)).

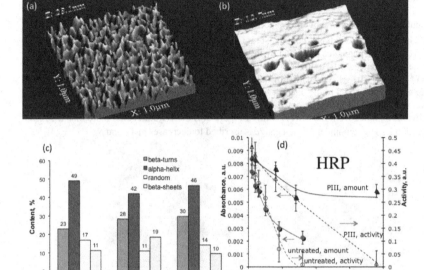

Figure 6: The AFM images show HRP protein layers on polystyrene films on silicon (scanned area is 1.0 x 1.0
mm). The coverage of protein is incomplete, with height indicative of aggregation, on the untreated surface (a – z
axis is 28.1 nm) and forms a densely packed monolayer on the PIII treated ($2.5e15$ ions/cm^2, N_2 20 kV) surface (b –
z axis is 12.7 nm). Ellipsometry and FTIR measurements indicate that there is a layer of protein with thickness ~ 9
nm on the surface as expected for a monolayer of HRP. The conformation (c), as determined by relative contents of
beta-turns, alpha-helices, random coils and beta-sheets in IR amide I peak is closer to native on PIII treated
polystyrene than when adsorbed on the untreated surface. (d) Amount (FTIR, solid symbols) and activity (TMB
assay, open symbols) of HRP protein are both higher on PIII treated than on untreated UHMWPE at all times of
storage in buffer.

Conclusions

We have developed a quantitative understanding of how unpaired electrons, mobile in a free
radical reservoir, are effective in covalent immobilisation of protein molecules. The dependences
of the rate of covalent immobilization on the age of the sample and the density of free radicals
show that the irreversible protein immobilization observed is associated with free radicals that
migrate from the a bulk reservoir to the surface.

Since they can be created on almost any material, such unpaired electron rich layers can be used as versatile interfaces for covalent coupling of functional biomolecules without the need for specific linker chemistry. Modulation of the chemistry of the environment to which the surfaces are exposed immediately after their creation allows tailoring of surface chemistry to optimise retention of protein function. Good covalent coupling will be achieved provided that the structures under the surface allow migration of the unpaired electrons to the surface on timescales compatible with the incubation time in protein solution. The immobilised protein layers on these surfaces are typically dense and close to void free. The conformation is closer to the native protein conformation on the hydrophilic ion treated surfaces than that of protein on untreated controls and the activity retention over time is significantly better due to a more robust attachment and a reduced degradation of enzyme activity. Such ion treated polymeric materials show promise for applications in biosensors and protein microarrays, microfluidics devices and for improving the biocompatibility of implantable biomedical devices.

Acknowledgements

We acknowledge the Australian Research Council for funding and the Wellcome Trust Equipment Fund for provision of the ESR. We acknowledge industry partners Cochlear Ltd and SpineCell Pty Ltd for financial and in-kind research support. We thank Professor Cristobal G. dos Remedios for advice on protein function and assays.

References

[1]   M. M. M. Bilek, *et al.*, "Free radical functionalization of surfaces to prevent adverse responses to biomedical devices," *Proc. Nat. Acad. Sci.*, p. in press, 2011.

[2]   W. F. DeGrado, "Computational Biology – Biosensor Design," *Nature*, vol. 423, pp. 132-133, 2003.

[3]   S. Hanash, "Disease proteomics," *Nature*, vol. 422, pp. 226-232, Mar 2003.

[4]   P. Olsson, *et al.*, "On the blood compatibility of end-point immobilized heparin," *J. Biomater. Sci. Polymer Edn.* , vol. 11, pp. 1261-1273, 2000.

[5]   C. Werner, *et al.*, "Current strategies towards hemocompatible coatings," *Journal of Materials Chemistry*, vol. 17, pp. 3376-3384, 2007.

[6]   C. Nojiri, *et al.*, "Nonthrombogenic Polymer Vascular Prosthesis," *Artificial Organs*, vol. 19, pp. 32-38, 1995.

[7]   C. P. Paweletz, *et al.*, "Reverse phase protein microarrays which capture disease progression show activation of pro-survival pathways at the cancer invasion front," *Oncogene* vol. 20, pp. 1981-1989, 2001.

[8]   W. Pompe, *et al.*, "Functionally graded materials for biomedical applications," *Materials Science and Engineering A*, vol. 362, pp. 40-60, 2003

[9]   M. Uchida, *et al.*, "Biomimetic coating of laminin-apatite composite on titanium metal and its excellent cell-adhesive properties," *Advanced Materials*, vol. 16, pp. 1071-1074, 2004.

[10]   D. V. Bax, *et al.*, "The linker-free covalent attachment of collagen to plasma immersion ion implantation treated polytetrafluoroethylene and subsequent cell-binding activity," *Biomaterials*, vol. 31, pp. 2526-2534, 2010.

[11]   D. Harman, "Aging: A Theory Based on Free Radical and Radiation Chemistry.," *J. Gerontol.* , vol. 11, pp. 298-300, 1956.

[12]   S. Giunta, *et al.*, "Transformation of beta-amyloid (AP) (1-42) tyrosine to L-Dopa as the result of in vitro hydroxyl radical attack," *Amvloid: Int. J. L h . Clin. Invest.* , vol. 7, pp. 189-193, 2000.

[13]   V. N. Popok, *et al.*, "High fluence ion beam modification of polymer surfaces: EPR and XPS studies," *Nuclear Instruments and Methods in Physics Research Section B: Beam Interactions with Materials and Atoms*, vol. 178, pp. 305-310 2001.

[14] B. J. Jones, *et al.*, "Electron paramagnetic resonance study of ion implantation induced defects in amorphous hydrogenated carbon," *Diamond and Related Materials*, vol. 10, pp. 993-997, 2001.

[15] G. Mesyats, *et al.*, "Adhesion of Polytetrafluorethylene modified by an ion beam," *Vacuum*, vol. 52, pp. 285-289, 1999.

[16] M. Bilek and D. R. McKenize, "Plasma modified surfaces for covalent immobilization of functional biomolecules in the absence of chemical linkers: towards better biosensors and a new generation of medical implants," *Biophysical Review*, vol. 2, pp. 55-65, 2010.

[17] Y. B. Yin, *et al.*, "Protein immobilization capacity and covalent binding coverage of pulsed plasma polymer surfaces," *Applied Surface Science*, vol. 256, pp. 4984-4989, 2010.

[18] Y. B. Yin, *et al.*, "Acetylene plasma polymerized surfaces for covalent immobilization of dense bioactive protein monolayers," *Surface & Coatings Technology*, vol. 203, pp. 1310-1316, 2009.

[19] Y. B. Yin, *et al.*, "Covalently Bound Biomimetic Layers on Plasma Polymers with Graded Metallic Interfaces for in vivo Implants," *Plasma Processes and Polymers*, vol. 6, pp. 658-666, 2009.

[20] Y. B. Yin, *et al.*, "Plasma Polymer Surfaces Compatible with a CMOS Process for Direct Covalent Enzyme Immobilization," *Plasma Processes and Polymers*, vol. 6, pp. 68-75, 2009.

[21] Y. B. Yin, *et al.*, "Covalent immobilisation of tropoelastin on a plasma deposited interface for enhancement of endothelialisation on metal surfaces," *Biomaterials*, vol. 30, pp. 1675-1681, 2009.

[22] Y. Yin, *et al.*, "Acetylene plasma coated surfaces for covalent immobilization of proteins," *Thin Solid Films*, vol. 517, pp. 5343-5346, 2009.

[23] N. J. Nosworthy, *et al.*, "The attachment of catalase and poly-L-lysine to plasma immersion ion implantation-treated polyethylene," *Acta Biomaterialia*, vol. 3, pp. 695-704, 2007.

[24] C. MacDonald, *et al.*, "Covalent attachment of functional protein to polymer surfaces: a novel one-step dry process," *Journal of the Royal Society Interface*, vol. 5, pp. 663-669, 2008.

[25] A. Kondyurin, *et al.*, "Attachment of horseradish peroxidase to polytetrafluorethylene (teflon) after plasma immersion ion implantation," *Acta Biomaterialia*, vol. 4, pp. 1218-1225, 2008.

[26] A. Kondyurin, *et al.*, "Covalent Attachment and Bioactivity of Horseradish Peroxidase on Plasma-Polymerized Hexane Coatings," *Plasma Processes and Polymers*, vol. 5, pp. 727-736, 2008.

[27] J. P. Y. Ho, *et al.*, "Plasma-treated polyethylene surfaces for improved binding of active protein," *Plasma Processes and Polymers*, vol. 4, pp. 583-590, 2007.

[28] D. V. Bax, *et al.*, "Linker-free covalent attachment of the extracellular matrix protein tropoelastin to a polymer surface for directed cell spreading," *Acta Biomaterialia*, vol. 5, pp. 3371-3381, 2009.

[29] S. L. Hirsh, *et al.*, "A Comparison of Covalent Immobilization and Physical Adsorption of a Cellulase Enzyme Mixture," *Langmuir*, vol. 26, pp. 14380-14388., 2010.

[30] R. Ganapathy, *et al.*, "Immobilization of papain on cold-plasma functionalized polyethylene and glass surfaces," *Journal of Biomaterial Science Polymer Edition* vol. 12, pp. 1027-1049, 2001.

[31] D. Kiaei, *et al.*, "Tight binding of albumin to glow discharge treated polymers," *Journal of Biomaterial Science, Polymer Edition*, vol. 4, pp. 35-44, 1992.

[32] M. M. M. Bilek, *et al.*, "Free radical functionalization of surfaces to prevent adverse responses to biomedical devices," *PNAS*, in press 2011.

[33] W. J. Wu, *et al.*, "Glycosaminoglycans mediate the coacervation of human tropoelastin through dominant charge interactions involving lysine side chains," *Journal of Biological Chemistry*, vol. 274, pp. 21719-21724, Jul 1999.

[34] B. K. Gan, *et al.*, "Etching and structural changes in nitrogen plasma immersion ion implanted polystyrene films," *Nuclear Instruments & Methods in Physics Research Section B-Beam Interactions with Materials and Atoms*, vol. 247, pp. 254-260, 2006.

[35] A. Kondyurin, *et al.*, "Etching and structural changes of polystyrene films during plasma immersion ion implantation from argon plasma," *Nuclear Instruments & Methods in Physics Research Section B-Beam Interactions with Materials and Atoms*, vol. 251, pp. 413-418, 2006.

[36] H. Jiang, *et al.*, "Surface oxygen in plasma polymerized films," *Journal of Materials Chemistry*, vol. 19, pp. 2234–2239, 2009.

Mater. Res. Soc. Symp. Proc. Vol. 1354 © 2011 Materials Research Society
DOI: 10.1557/opl.2011.1471

Neural Cell Attachment on Metal Ion Implanted Glass Surfaces

Emel Sokullu-Urkac,[1*] Ahmet Oztarhan,[1,2] Ismet Deliloglu-Gurhan,[2] Sultan Gulce-Iz,[2] Feyzan Ozdal-Kurt[3] and Ian G. Brown[4]
[1]Surface Modification Laboratory, Ege University, 35100 Izmir, Turkey
[2]Bioengineering Department, Ege University, 35100 Izmir, Turkey
[3]Biology Department, Celal Bayar University, 45040 Manisa, Turkey
[4]Lawrence Berkeley National Laboratory, Berkeley, CA 94720, USA

ABSTRACT

We have explored the application of ion implantation as a tool for the enhancement of neural cell growth on glass surfaces. Glass substrates were ion implanted with gold and with carbon using a metal vapor vacuum arc (MEVVA) ion source-based implantation system at Ege University Surface Modification Laboratory. The implantation dose was varied over the range $10^{14} - 10^{17}$ ions/cm^2 and the ion energy spanned the range 20 – 80 keV. B35 neural cells were seeded and incubated on the implanted substrates for 48h at 37°C. After 2-days in culture the cell attachment behavior was characterized using phase contrast microscopy. The adhesion and direct contact of neural cells on these ion implanted glass surfaces were observed.

INTRODUCTION

There is potential for the use of ion beam modification of surfaces for basic neurobiological research, with a variety of possibilities lying at the interface between inorganic and organic neuron systems, such as pain relief through neuron stimulation, regeneration of defective neurons, artificial retinas, and the realization of true neural computing [1-3]. Biocompatibility of the implant or substrate material is an important factor. The assessment of biocompatibility involves the ability of tissue to adhere to the surface of the material, and in general the more compatible the surface the greater the cell adhesion.

In the work described here we have concentrated on the viability and morphology of neural cells on surfaces modified by ion implantation. The ion implantation was carried out over a range of implantation dose and ion energy, and cell culture experiments were done on all samples under the same condition. In this way, we were able to compare the cell attachment behavior on the modified surfaces.

Our study covers cell attachment and network formation of neurons on Au- and C-implanted glass surfaces. B35 neuroblastoma cell line was used for in vitro studies. B35 cells have proven useful model in the molecular analysis of endocytosis and of signaling pathways and in particular those that guide axonal outgrowth and cell motility [4]. In this study, we examined growth of cells in number in a culture medium with serum on ion implanted and unimplanted samples. Thus we examined the effect of ion implantation on neural cell adhesion and axonal outgrowth.

EXPERIMENTAL DETAILS

We used glass rather than polymer as the substrate material in order to eliminate any effects of polymeric interactions on the surfaces.

Surface modification

Glass substrates (microscope slides) were ion implanted using a vacuum arc ion source-based ion implantation system at the Ege University Surface Modification Laboratory. This facility has been described in detail elsewhere [5]. Prior to implantation all glass samples were cleaned in an ultrasonic bath for 10 min. in deionized water and then rinsed with 70% ethanol. Sterilized samples were ion implanted with carbon (C) and gold (Au), separately. Implantation was done with doses of $10^{14} - 10^{17}$ and ion energy 20 – 80 keV. After surface modification, all samples were sterilized with autoclave for 45 min.

Cell viability

B35 cells (ATTC, CRL2754) were seeded at a cell density of 4×10^5 cell/ml and incubated in a humid, 5% CO_2 cell culture incubator at 37°C for 48h. The cell culture medium was DMEM, Gibco supplemented with 10%FBS, 1% L-glutamin, 100 U/ml penicillin, and 100 mg/ml streptomycin. If otherwise specified, all cell culture reagents were purchased from Biochrom, Germany. After 2-days in culture, the attachment of neural cells on the modified surfaces was observed using inverted phase contrast microscopy (Olympus CK-40, Japan). Cell contrast ratio was calculated as the ratio of number of cells attached on the implanted sample to that on an untreated sample.

$$\text{Cell Contrast Ratio} = \frac{\text{\# of cells attached on ion implanted surface } (cell/area)}{\text{\# of cells attached on pristine surface } (cell/area)}$$

RESULTS and DISCUSSION

Microscope images of B35 neural cell attachment on C- and Au-implanted glass surfaces are shown in Figures 1 and 2, respectively; these images represent results spanning the complete range of ion beam parameters used. We point out that in vacuum arc ion sources the ions are in general multiply ionized. In particular, while carbon is singly ionized as C^+, gold is primarily doubly ionized as Au^{2+} [6,7]. Thus ion source extraction voltages of 20 – 40 kV correspond to mean ion energies of 20 – 40 keV for C^+ and 40 – 80 keV for Au^{2+}. Cell contrast ratio results for each parameter combination (in terms of dose and energy) for Au-ion implantation and C-ion implantation are shown in Table 1 and Figure 3.

Figures 1 and 2 show the morphology of the neural cells cultured on the C- and Au-implanted glass surfaces. It can be seen that the cell adherence is greater on surfaces treated at higher implantation dose and energy. In Figure 1, axon outgrowth is best for C implantation at 30 keV energy and 1×10^{16} dose. A similar trend shows in Figure 2 for Au implantation, with best result observed at 60 keV ion energy and 10^{15} dose. However, in the lower dose range in both Figures 1 and 2, no whole cells with good morphology are found. We conclude that the most suitable parameter set for Au implantation is 60 keV energy and 1×10^{15} ion/cm^2 dose, and for C implantation 30 keV energy and 1×10^{16} ion/cm^2 dose. Table 1 and Figure 3 show that improvement in cell contrast ratio is nearly a factor of 3.5 for the above conditions, for both C- and Au-implantations.

In Figure 2, for Au implantation at a dose of 1×10^{15} ions/cm^2, while the best cell attachment results were for 60 keV, the 80 keV ion energy samples show similar results. Both Figures 1 and 2 show neurons connected with each other in a network, especially for energies

16

greater than 30 keV (C) and 60 keV (Au). Neural cells are adhesive contact dependent, which is a prerequisite for surviving and for neurite outgrowth [8,9]. Overall we can say that the outgrowth of axons on the ion beam modified glass surfaces shown in Figures 1 and 2 indicate good adherence to the surface.

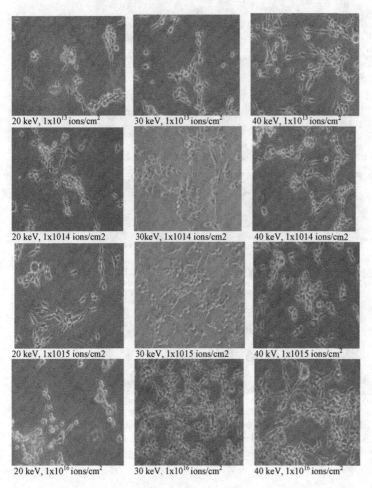

20 keV, 1×10^{13} ions/cm^2 30 keV, 1×10^{13} ions/cm^2 40 keV, 1×10^{13} ions/cm^2

20 keV, 1x1014 ions/cm2   30keV, 1x1014 ions/cm2   40 keV, 1x1014 ions/cm2

20 keV, 1x1015 ions/cm2   30 keV, 1x1015 ions/cm2   40 kV, 1x1015 ions/cm^2

20 keV, 1×10^{16} ions/cm^2 30 keV. 1×10^{16} ions/cm^2 40 keV, 1×10^{16} ions/cm^2

Figure 1. Neural cell (B35) attachment (48h) on C-implanted glass surfaces for different dose and energy values. The implantation dose and energy is shown at the bottom of each image. (Phase contrast microscope, 20X , Olympus, Japan).

17

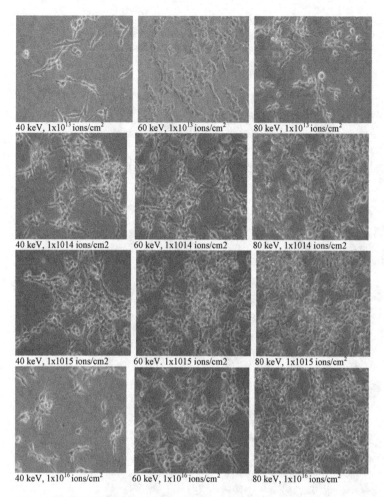

Figure 2. Neural cell (B35) attachment (48h) on Au-implanted glass surfaces for different dose and energy values. The implantation dose and energy is shown at the bottom of each image. (Phase contrast microscope, 20X , Olympus, Japan).

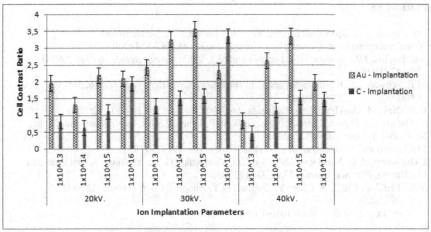

Figure 3. Cell contrast ratio (with standart deviation) for neural cell (B35) attachment (48h) on C- and Au-implanted glass surfaces for different dose and energy values. (Extraction voltage in kV and dose in ions /cm^2).

Table I. Cell contrast ratio for neural cell (B35) attachment (48h) on C- and Au-implanted glass surfaces for different dose and energy values. (Extraction voltage in kV and dose in ions/cm^2).

		Extraction Voltage					
		20kV.		30kV.		40kV.	
		Au	C	Au	C	Au	C
Dose (ion/cm^2)	1×10^{13}	1,9	0,8	2,4	1,3	0,8	0,5
	1×10^{14}	1,3	0,6	3,2	1,5	2,6	1,1
	1×10^{15}	2,2	1,1	3,6	1,6	3,3	1,5
	1×10^{16}	2,1	1,9	2,3	3,4	1,9	1,4

CONCLUSION

The work described here has shown that neural cells grown on ion implanted glass surfaces display tight adherence and good contact with the surface, a result that bodes well for future possible neural network applications. The neural cell adhesion and axonal outgrowth changes according to the ion beam parameters (dose and energy) used for the implantation. C- and Au-implantation of glass surfaces at lower energies can provide a tool for enhancing neural interfaces.

ACKNOWLEDGMENTS

This research was supported by the TUBITAK (Scientific & Technological Research Council of Turkey) Research Grant No. 108M391.

REFERENCES

1. I.G. Brown, K.A. Bjornstad, E.A. Blakely, J.E. Galvin, O.R. Monteiro and S. Sangyuenyongpipat, *Plasma Phys. Control. Fusion* **45**, 547-554 (2003).
2. S.C. Bayliss, L.D. Buckberry, I. Fletcher and M.J. Tobin, *Sensor Actuat. A-Phys.* **74/1-3**, 139-142 (1999).
3. M.E. Schwab, J.P. Kapfhammer and C.E. Bandtlow, *Annu. Rev. Neurosci.* **16**, 565-595, (1993).
4. C.A. Otey, M. Boukhelifa and P. Maness, *Method. Cell Biol.* **71**, 287-304, (2003).
5. A. Oztarhan, I. Brown, C. Bakkaloglu, G. Watt, P. Evans, E. Oks, A. Nikolaev and Z. Tek, *Surf. Coat. Technol.*, **196**, 327 (2005).
6. I.G. Brown and X. Godechot, *IEEE Trans. Plasma Sci.* **19**, 713 (1991).
7. I. Gushenets, A.G. Nikolaev, E.M. Oks, L.G. Vintizenko, G.Yu. Yushkov, A. Oztarhan and I.G. Brown, *Rev. Sci. Instrum.* **77**, 063301 (2006).
8. Y.W. Fan, F.Z. Cui, L.N. Chen, Y. Zhai and Q.Y. Xu, *Surf. Coat. Technol.* **131/1-3**, 355-359 (2000).
9. H. Rauvala, *J. Cell Biol.* **98:3**, 1010-1016 (1984).

Mater. Res. Soc. Symp. Proc. Vol. 1354 © 2011 Materials Research Society
DOI: 10.1557/opl.2011.1081

Cluster Ion Beam Processing: Review of Current and Prospective Applications

Isao Yamada[1] and Joseph Khoury[2]
[1] Graduate School of Engineering, University of Hyogo, 2167 Shosha, Himeji, 671-2280 Japan.
[2] Exogenesis Corporation, 20 Fortune Drive, Billerica, Massachusetts 01821 USA

ABSTRACT

Cluster ion beam processes which employ ions comprised of a few hundred to several thousand atoms are being developed into a new field of ion beam technology. The processes are characterized by low energy surface interaction effects, lateral sputtering phenomena and high-rate chemical reaction effects. This paper reviews the current status of studies of the fundamental cluster ion beam characteristics as they apply to nanoscale processing and present industrial applications. As new prospective applications, techniques are now being developed to employ cluster ions in surface analysis tools such as XPS and SIMS and to modify surfaces of bio-materials. Results related to these new projects will also be reviewed.

INTRODUCTION

In cluster ion beam bombardment of solid surfaces, the concurrent energetic interactions between many atoms comprising a cluster and many atoms at a target surface result in highly non-linear sputtering and implantation effects. Following successful accomplishment in 1988 of intense beams of gas clusters from small nozzles at room temperature, an extended series of investigations was conducted at Kyoto University and at the University of Hyogo to develop gas cluster ion beam (GCIB) fundamentals and applications [1].

By 1995, it was recognized that it would not be practical to use substantially greater gas flows in order to increase cluster ion beam currents to the levels to be required for production processors. No other groups or institutes in the world had yet paid attention to the concept of cluster ion beam equipment and to possible uses of gas cluster ions for surface processing. In collaboration with the author's group (IY) at Kyoto University, in support of work sponsored by the Japan Science and Technology Agency (JST), Epion Corporation in the US began development of commercial GCIB equipment in 1995 [2]. Efforts to increase cluster generation, to improve efficiency of cluster ionization, and to optimize beam transport without increasing gas consumption or pumping requirements, were successful. Cluster ion beam currents of several hundred microamperes on target became possible with source gas flows that could be handled by standard vacuum pumps. Commercial GCIB equipment by Epion was introduced in 2000.

Historically, cluster ion beam processing efforts expanded into two different but closely related major categories: gas cluster ion beam (GCIB) processing and polyatomic ion beam processing. GCIB has become useful in a number of nanotechnology areas by employing its unique characteristics of very low effective energies, its lateral sputtering effects and its high chemical reactivity effects. Polyatomic ion beam technology investigations were initially started during the early GCIB research in order to experimentally demonstrate the low energy interaction effects which are associated with bombardment by multiple atom particles. Because a GCIB beam contains a wide range of cluster sizes, typically from a few hundred atoms to many thousands of atoms, it had during early investigations been difficult to quantitatively describe the dependence of low energy interaction effects upon cluster size. In order to obtain clear experimental evidence, it was desirable to use cluster ions of a single specific number of atoms per cluster. The idea then emerged to use a molecular ion consisting of a relatively large number of primary atoms of one single element. The polyatomic material candidate selected was decaborane ($B_{10}H_{14}$) which consists of 10 boron atoms bonded together by low mass hydrogen atoms. Note that a GCIB system equipped with a cluster size selection system had not yet been developed at that time. As the research proceeded, experimental implantation of decaborane ions into Si showed clearly that the effective energy of each boron atom of a polyatomic cluster ion is essentially equal to the total energy of the cluster divided by the number of boron atoms contained in the cluster. As the research continued, successful fabrication of a P-MOSFET with 40 nm gate was accomplished for the first time in 1996 [3].

In order to develop commercial polyatomic cluster ion implantation equipment, the author (IY) arranged JST sponsorship for a "risk-taking technological development project" (in Japanese, Itaku-Kaihatsu) which started in 1998 at Sumitomo Eaton Nova. The program was selected from among candidates involving JST held patents which originated from university research projects where patented technology was considered to have high probability of being successfully developed for practical use and for promoting further research. The risk-taking contract ended in 2001, but the technology was not yet fully developed and serious concerns still existed regarding whether the aggregate atoms from molecular clusters could be adequately annealed into substitutional sites and whether end-of-range effects and transient enhanced diffusion issues could be overcome. Further collaboration was subsequently made between Kyoto University, Nissin Ion Equipment Co., Ltd. and Fujitsu under JST support. This project was started in 2003 and was successfully completed in 2005. The resulting polyatomic ion implantation concept has become a major low energy processing method for ultra-shallow junction formation.

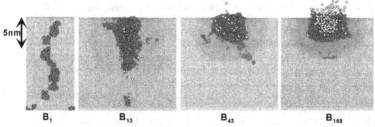

Figure 1. MD simulation results of atomic and cluster Boron ion implantation into Silicon at 7keV after 385fs.

Over the 20 years of cluster ion beam investigations, low energy surface interaction effects, lateral sputtering phenomena and high-rate chemical reaction effects were explored experimentally and were explained by means of molecular dynamics (MD) modeling [4]. The fundamental results concerning ion/solid surface interactions have become useful and valuable information and they have become a base for a considerable amount of application technology. Cluster ion beams which have such distinctive characteristics have been applied for nanoscale processing such as shallow junction formation, low damage surface modification and etching, ultra-smooth surface formation and high quality thin film formation.

A fundamental question in cluster ion interactions, as compared to interactions by traditional atomic and molecular ions, was how large must a cluster ion be in order for it to produce non-linear effects during bombardment of a solid surface? The answer became clear when theoretical modeling done in 1996 by Takaaki Aoki was able to correctly predict the experimental observations [5]. Figure 1 shows MD modeling snapshots of monomer (B_1), small cluster (B_{13}) medium cluster (B_{43}) and large cluster (B_{169}) ions impacting at 7 keV onto crystalline Si at an elapsed time of 385fs following impact. The results of these simulations have shown that cluster-like bombardment phenomena, ie. the nonlinear effects which are typical of cluster impact, are already evident at a size of 13 atoms. The result for B_{13} does exhibit concentration of displacement damage, but damage from larger gas clusters is seen to be much more concentrated and more localized. It was predicted that complete self-amorphization resulting with cluster impact could help cause better solid phase epitaxy during low temperature annealing.

Recent successful industrial applications of GCIB are for fabrication of thin film transistors (TFT) using the low energy effect [6], for local corrective etching in bulk acoustic wave devices using the lateral sputtering effect [7] and for surface planarization of patterned hard disk drive (HDD) media using the low energy and lateral sputtering effects [8]. In these applications, nanoscale dimensional precision has been achieved, for example in formation of shallow junctions of depths less than a few nm, and in smoothing of surfaces to roughness less than 1 nm. Figure 2 illustrates some of the milestones of GCIB equipment and process development.

New GCIB applications are now being developed, for example in shallow surface analysis

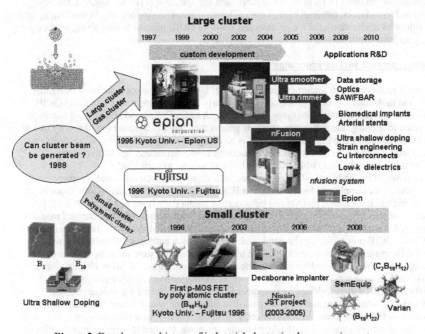

Figure 2. Development history of industrial cluster ion beam equipment

techniques such as XPS and SIMS. Because a large cluster can sputter from a much shallower volume of material than in the case of a smaller projectile, GCIB can be beneficially employed to accomplish extremely surface-sensitive analyses of organic thin films. Another new area of application is in surface modification of bio-materials. It has been shown by Exogenesis Corporation in the US that GCIB modification of medical device materials such as titanium can greatly enhance osteoblast proliferation and bone formation. Exogenesis has also demonstrated a novel method of drug delivery from metal vascular stents in which GCIB is employed to control rate of drug elution without requiring the use of any binding polymer.

This paper reviews some of the present industrial nano-scale applications of GCIB and discusses new results related to surface modification of bio-materials.

APPLICATIONS TO NANOSCALE PROCESSING

Ability to modify ultra-thin surfaces under extremely low damage conditions and with nanometer precision has become important in recent device fabrication. The METI (Japanese Ministry of Economy and Trade and Industry) R&D Program & Technology Strategy Map has been published as a tool for strategic planning and implementation of R & D investment in several industrial areas [9]. METI roadmaps published in 2007 through 2010 have recommended cluster ion beam processing for uses in the nanotechnology fields of next-generation semiconductor devices (TFT's of Si, SiC, GaN), data storage devices (HDD, MRAM, MEMS), sensors and transducers, extremely small optical lenses, photonics devices, etc. In working upon such industrial applications, the capabilities and limits of GCIB relative to smoothing, surface damage and nm-depth etching have become recognized.

Extremely smooth surface processing

Lateral sputtering is an important characteristic of GCIB which is not associated with other atomic and molecular ion beam processes. During early investigations it was found that GCIB can produce surface smoothing of many materials, such as CVD diamond, glasses and metals, even when its etching rate in these materials is greater than that of monomer ion beams by two to three orders of magnitude [2]. Because GCIB is an energetic beam process, it was reasonable to expect that it might not be able to produce smoothing to sub-nanometer levels, but on the contrary it has been shown to offer excellent ability to smooth to such levels.

The limits of the ability of GCIB for surface smoothing have been evaluated for many applications, for example, in planarization of a HDD surface. In order to increase the capacity of a HDD, discrete track media (DTM) and bit patterned media (BPM) are being developed [8]. However, for these approaches, nano-level smoothing and planarization of the patterned media are required. To utilize GCIB to achieve such accurate nano-level surfaces, the lateral sputtering behavior of GCIB processing can be employed. For evaluation purposes, test substrates with several different line-and-space patterns were used. The substrates were made by nano-imprint lithography on CVD-deposited TiCr films. The substrate groove intervals were 100 nm and the depth of each groove was 18 nm. Figure 3 shows AFM images of the surface patterns after irradiation by GCIB at 20 keV and by monomer ions at 400 eV. Before the irradiations, the roughness, Ra, was 2.47 nm and the P-V (peak-to-valley) value was 4.75 nm. After the GCIB irradiation, surface roughness Ra was reduced to 0.46 nm and P-V value was reduced to 0.50 nm. The identical sample irradiated by the Ar monomer ion beam resulted in surface roughness Ra of 0.9 nm and P-V of 2.88 nm. This example exhibits the capability of GCIB irradiation for decreasing surface roughness even in the sub-nm region. In order to investigate the dimensional range characteristics of smoothing effects produced by GCIB, atomic force microscope (AFM) examinations have been made on Ar-GCIB bombarded amorphous carbon film line & space patterns with groove intervals from 100 to 400 nm [10]. Spatial wavelength dependence of the reduction in roughness caused by the GCIB exposures was studied using the power spectrum of the AFM images. As is shown in figure 4, the reduction of power spectrum intensity decreased monotonically with decreasing groove interval and was particularly remarkable at spacings below 200 nm. Because surface planarization by GCIB is dominated by the lateral motion of atoms induced by individual cluster ion impacts, structures with very small spatial dimensions are removed preferentially.

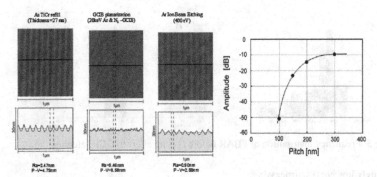

Figure 3. AFM images of surfaces before irradiation, after GCIB irradiation and Ar monomer ion irradiation.

Figure 4. Power spectra of AFM images after GCIB irradiation.

Precise nanometer depth etch processing

Precise nm-depth etching utilizing the low energy effects of GCIB bombardment is becoming important in various industrial applications. To achieve such etching, the inherent GCIB character which combines a low energy bombardment process with the above mentioned lateral sputtering behavior seems to be unique among existing technologies. In conjunction with its low energy effects, GCIB can deliver a highly directional and controllable energetic chemical beam. These characteristics have been employed by Epion Corporation in several distinct nm-processing areas, for example in fabrication of Film Bulk Acoustic Resonator (FBAR) filters, EUV masks, OLED (Organic Light Emitting Diode) displays, etc [11]. An important approach to applying GCIB processes has been incorporation of a Location Specific Processing (LSP) control system which utilizes a scanning capability in which the dwell time is varied as a function of the amount of material to be removed at any specific location. A wafer metrology map is imported into the LSP control system to define the functional relationship between x,y position on the wafer and required etching depth. LSP is now being utilized in several surface smoothing applications such as production of large Si wafers, MEMS, and small molds for optical lens fabrication, etc.

Through cooperation between Epion Corporation in the US and EPCOS AG in Munich,

Germany, FBAR processing using GCIB-LSP has been successfully developed. Operating frequencies of FBAR filter devices depend very critically upon thickness of a piezoelectric material (AlN) surface layer and adequate frequency uniformity across individual wafers and from wafer to wafer is difficult to achieve. In order to achieve increased production yields, GCIB-LSP is used to remove precisely controlled amounts of surface material in patterns which are adjusted for each individual wafer [7]. As shown in figure 5, a reduction of width of frequency distribution by factor of 5 over a typical 200 mm substrate area has been reported. Piezoelectric quality of the highly textured, polycrystalline, c-axis oriented columnar surface film is not affected by the GCIB exposure.

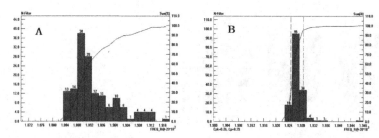

Figure 5. Frequency distribution of FBAR before (a) and after (b) GCIB etching.

Extremely low energy processing

After the characteristic low energy behaviors of GCIB were first experimentally demonstrated in 1993 [12], GCIB received attention from the nanotechnology community for various potential uses in semiconductor, magnetic, and optoelectronic devices. In addition to its characteristics for surface processing GCIB also became of interest for its secondary electron and ion emission characteristics.

In 1999, N. Toyoda described the dependence upon bombarding ion velocity of secondary ion yields from Au surfaces impacted by C cluster ions of size n (n=1, 7, 9, 11, 60) [13]. Figure 6 shows the reported secondary ion yields normalized to cluster size, i.e., the number of secondary ions produced per cluster atom, versus velocity of the clusters. For cluster sizes up to 7 atoms, the yields are seen to be almost same as that produced by the carbon monomer ions. However, an enhanced secondary ion yield was observed starting from cluster size of 9. These results were adopted for SIMS analysis for the first time by the author's group at Kyoto University in 2001 [14]. It was expected that high resolution depth profiling could be achieved because of the high sputtering yields and low energy irradiation effects in combination with minimal ion mixing behavior and absence of surface roughening. The experiments demonstrated a resolution of several nm achieved using Ar cluster ions as a primary ion beam.

Further development of GCIB TOF-SIMS was subsequently made by the J.Matsuo group at Kyoto University [15]. Figure 7 shows a schematic of a GCIB TOF-SIMS. A primary Ar cluster ion beam is size-selected by a double deflection method, and is then introduced upon the sample. Secondary ions produced by GCIB are accelerated to a kinetic energy of 2 keV and are detected with a microchannel plate (MCP). The timing of the secondary-ion chopping and the detection are used as the start and stop signals for the time-of-flight (TOF) measurement. SIMS depth profiling of ITO/PS, Si/DSPC and ITO/Alq3/NPD samples showed that the intensities of the molecular ions from these samples remained constant with increasing fluence. Stable intensities at steady state and at least two orders of magnitude of dynamic range have been reported and depth resolution of SIMS profiling with Ar cluster ion beams is estimated to be better than 10 nm [16]. These results have confirmed the potential for depth profiling analyses using beams of large Ar cluster ions.

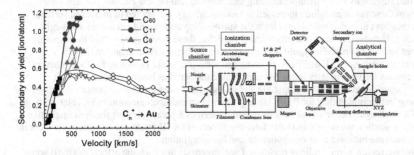

Figure 6. Secondary ion emission yields by carbon cluster ions form Au surface.

Figure 7. Schematic of GCIB TOF-SIMS

A size-selected GCIB SIMS developed by K.Mochiji and colleagues at University of Hyogo has emphasized damageless processes for large organic materials [17]. In organic materials evaluation by conventional secondary ion mass spectrometry (SIMS), the molecular weight of the intact ions currently detectable is at best only as high as 1000, which prevents the technique from being applied to biomaterials of higher mass. However the developed GCIB-SIMS could detect intact ions of insulin (molecular weight: 5808) and cytochrome C (MW: 12327). The results indicate that fragmentation could be substantially suppressed without sacrificing the sputter yield of intact ions when the kinetic energy per atom was decreased to the level of the target's dissociation energy.

Ar GCIB gas has been applied for low damage depth profiling by XPS and for removal of damaged layers of polymer materials. ULVAC-Phi Inc. announced commercial availability of GCIB mounted XPS equipment in 2011 [18]. XPS depth profiling of polyimide thin films using GCIB have shown the extremely low damage during the depth profiles, compared with the depth profiles obtained by C_{60} and coronene ion sputtering.

APPLICATIONS TO BIO-MATERIALS

Surface Integration using GCIB

It is well understood that for a cell to attach to a surface which will eventually lead to proliferation and differentiation depends on the wettability or hydrophilicity, the chemistry, the charge, and the roughness. Surface biocompatibility and cytocompatibility, therefore, are generally regarded as a need for integration of a device into the human body. Ranging from conductive metallic implants such as used in the dental and orthopedic fields, to insulating polymer implants such as polyester or polytetrafluoroethylene (PTFE, Teflon®) vascular grafts and polyether ether ketone (PEEK) spinal cages, to allograft tissues such as anterior cruciate ligament (ACL) repair or bone grafts, the more biocompatible a surface chemistry is, the less the body will react negatively towards it, and the more cytocompatible a surface is, the better it will integrate with the surrounding tissue. Many advances have been made in various surfaces over the last few decades. In the case of titanium, which is considered to be biocompatible, implants used in dentistry have gradually progressed from a smooth surface to a machined surface to a roughened surface by many means including sandblasting and/or acid etching [19, 20]. The changes in these surfaces, which increased cytocompatibility, merely led to increased surface area or roughness in which osteoblasts from the surrounding bone could latch onto and begin integration. However, at least one group has argued that these modifications lead to temporary increased surface hydrophilicity and hence the reason for better cell attachment [21].

Materials such as PTFE or PEEK, to name just a couple, while favorable for their inherent characteristics for various uses in medicine and are considered as biocompatible, are not generally considered to be cytocompatible and as such have poor integration potential. Increasing cytocompatibility for PTFE leading to rapid re-endothelialization in vascular grafts or for PEEK leading to better spinal fusions would be seen as a paradigm shift for their respective uses. In our studies, we sought to understand the effects of GCIB treatment of various surfaces in order to increase bio- and cyto-compatibility and bio-integration.

Here, we review some findings performed at Exogenesis in which the effects of GCIB were analyzed in respect to enhancement of cell adhesion, proliferation, and differentiation allowing better integration of a surface into the body.

GCIB modifications of surfaces of implantable medical devices

Due to its inert properties, argon gas (Ar) was selected to produce the clusters in which to modify the surface. Ar clusters dissipate upon impact with the surface, leaving no residue behind; therefore, nothing is added onto or into the surface. A wide range of doses as determined by clusters per cm^2 were studied, here we present findings using $5x10^{14}$ ions/cm^2 unless otherwise noted. At first we sought to understand the physical characteristics of the GCIB- modified surfaces. Surfaces become more hydrophilic following GCIB-treatment. The hydrophilicity of a surface is recognized to be important for initial cell attachment [21]. The hydrophilic properties, however, vary from surface to surface. The effect of enhanced hydrophilicity on titanium is relatively short lived, lasting from 3 to 48 hours (Figure 8a). The effect on PTFE, on the other hand, lasts much longer. We have patterned a sheet of PTFE using a mask and the GCIB-treated pattern has remained hydrophilic for over two years so far, and still counting (Figure 8b). The next change that we describe matches an earlier finding showing amorphization of surface crystallinity [22]; amorphous surfaces have been shown to be beneficial for cell proliferation [23, 24]. Earlier work also previously demonstrated the effect of the cluster bombardment on the formation of surface craters [25]. We believe that for bio-medical purposes these nano-craters lead to the formation of nano-roughness on the initial micro-roughness of the surface, thereby creating more surface area, and as described earlier, increased surface area leads to better cell growth. Finally, a modification that is currently being studied is change in surface charge potential. Since the Ar clusters that bombard the surface are charged, we believe that at impact charges transferred to the surface can cause beneficial alteration of surface charge potentials. Although the exact mechanism for enhanced cell attachment and proliferation on GCIB-treated surfaces has not been elucidated, a combination of enhanced hydrophilicity, crystalline amorphization of the surface, nanoscale roughness, and changes in surface charge potential are responsible for enhanced cytocompatibility.

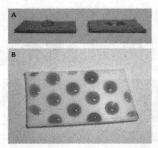

Figure 8. GCIB enhances hydrophilic properties of surfaces. A: shows a 5µl drop of water on titanium control surface (left) and GCIB treated surface (right). B: colored water remains on the GCIB-patterned surface of PTFE.

GCIB modification of surfaces leads to enhanced cell attachment and proliferation

Vascular grafts using PTFE or polyethylene terephthalate (PET, Dacron®) is commonplace; the idea to use PTFE or PET, very hydrophobic surfaces, is that cells do not bind to the surface and decrease the incidence of thrombosis. However, smaller-diameter grafts (<6mm) display an increased level of in-graft thrombosis [26, 27]. Rapid re-endothelialization of the graft lumen would be the ideal anti-thrombogenic surface. Many groups have attempted to address this issue by modifying PTFE or PET by adding Arg-Gly-Asp (RGD) attachment peptides, coating with adhesive proteins such as fibronectin or collagen, gas plasma treatment, or by chemically modifying the surface [28]. Such modifications, however, either do not last long term in the case of plasma treatment [29], or are not favorable due to potential changes in biocompatibility. Our studies have shown that GCIB treatment of both PTFE and PET leads to significant cell attachment and proliferation as compared to controls (Figure 9a-e). Further, these surface changes appear to be permanent as evidenced both by wettability of PTFE as shown in Figure 8b and in shelf life studies where the significant increase in cell attachment and proliferation were observed one year after GCIB treatment.

These changes have not only been demonstrated on PTFE and PET, but also successfully on titanium, polystyrene, glass, sapphire, PEEK, tissues derived from the body such as bone, ligament, and tendons, and many other surfaces. Enhancing cytocompatibility of these materials could lead to better integration of implantable medical devices and allografts or to better cell adhesion and growth for stem cell research. Studies on titanium have yielded very favorable results in significantly increased cell proliferation by day 10 time point 74% increase of cell number on GCIB as compared to controls (p<0.03) (Figure 10a-b). This increase could result in GCIB treatment of dental or orthopedic implants which osseointegrate in significantly decreased times. Similarly, PEEK which is now commonly used in spinal applications is preferred over titanium due to its elastic and compression properties more closely matching bone, however, due to its poor cytocompatibility, results in intra-body fusions that do not integrate properly. In initial studies, we have demonstrated that GCIB-treated PEEK result in cell attachment and proliferation comparable to native titanium (Figure 10c). Increasing the cytocompatibility of PEEK by GCIB would allow for inter-body fusions to occur without the addition of materials such as Bone Morphogenic Protein-2 (BMP-2) or demineralized bone powder.

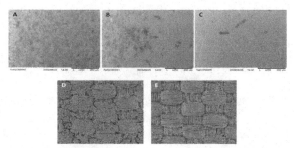

Figure 9. GCIB enhances cell attachment on surfaces. Osteoblast cells adhere and proliferate on PTFE that has been GCIB-treated (A) while no apparent cell attachment is seen on control PTFE (C). The portion that is half-masked (B) displays the difference in cell attachment on the surface of GCIB-treated PTFE. Endothelial cells seeded onto PET for 24 hours attach significantly more to GCIB-treated surfaces (D) as compared to control (E).

Enhancing cell proliferation alone is not indicative of better integration into the body if the proper differentiation of the cells to the desired tissue is not observed. In order to determine if the cells growing on the GCIB-treated surfaces are indicative of enhanced integration into the body, we looked at osteoblasts growing on titanium surfaces to verify if they are differentiating along the requisite pathway for bone development. An early indication of differentiation of the osteoblast cells is the increase in messenger RNA (mRNA) of various genes known to be involved in bone formation. Up-regulation of the ALPL gene is known to be involved in matrix mineralization leading to bone formation. Osteoblast cells growing on GCIB-treated titanium surfaces display 3.4 fold increase in ALPL gene as compared to cells growing on control surfaces by qPCR at 10 days (p<0.02). To verify the increase of the ALPL gene, we assayed mineralization by a well known method applying Alizarin Red to quantify the level of mineralization produced by osteoblasts growing on GCIB-treated or control titanium. Osteoblasts growing on GCIB-treated titanium produced 2.2 fold increase in mineralization as compared to growing on control surfaces, this is indicative of the osteoblasts undergoing osteogenesis and producing bone tissue.

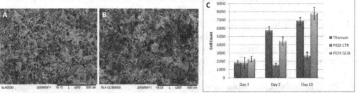

Figure 10. GCIB GCIB enhances cell attachment and proliferation. Titanium surfaces that were left as controls (A) did not allow osteoblast proliferation to occur to the same extent as GCIB-treated surfaces (B) by day 10. Analysis of the cell proliferation on PEEK surfaces shows GCIB-treated PEEK have a similar ability to allow osteoblasts to grow on the surface as untreated titanium (C).

In order to show proof of concept that GCIB surface modification leads to better integration, we did an *in vivo* study using a rat calvarial critical size defect model to show bone growth on

30

the surface of GCIB-treated and –untreated PEEK. Following 4 weeks after implantation into the calvarial defect, histology was performed to determine the amount of bone re-growth on the surface. It was found that GCIB-treated PEEK resulted in a bone ledge growing on top of the disk covering approximately 50% of the surface whereas the control PEEK resulted in no bone growth at all (Figure 11).

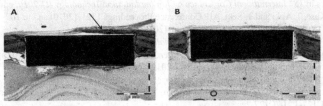

Figure 11. GCIB-treated PEEK results in bone formation in an in vivo model. GCIB-treated PEEK disks (A) result in significant bone coverage as indicated by the arrow as compared with control disks (B) in a four week study using a rat calvarial defect model.

CONCLUSIONS

GCIB processes involve low energy interaction effects, lateral sputtering behavior and high rate chemical reaction phenomena which are distinctly different from the characteristics exhibited by other types of atomic and molecular ion beam processes. The unique capabilities of GCIB are being utilized in surface smoothing, nm-depth etching and low-damage surface processes. Examples which have been described, including planarization of HDD media surfaces, fabrication of film bulk acoustic resonator filters, and low-damage depth profiling in XPS/SIMS analyses, demonstrate that GCIB can be employed for precise nanometer level processing in a broad range of industrial applications.

Other new prospective applications of GCIB in the processing of bio-materials have been introduced. GCIB shows potential to be used to enhance integration of implantable medical devices into the body by allowing better cell attachment, proliferation, and differentiation of surrounding tissue.

ACKNOWLEDGMENTS

We wish to thank Allen Kirkpatrick for his dedication to the development and understanding of GCIB technology, and for his critical review and help in the preparation of this manuscript. We also would like to thank Sean Kirkpatrick, Son Chau, and Raymond Cherian for their technical help with the biological studies.

REFERENCES

1. I. Yamada, in *18th International Conference on Ion Implantation Technology IIT 2010*, edited. by J. Matsuo, M. Kase, T. Aoki and T. Seki, AIP Conference Proceedings 1321, (2010) pp.1-8.
2. I. Yamada, J. Matsuo, N. Toyoda, and A. Kirkpatrick, Materials Science and Engineering, R34, (2001) pp.231-295.
3. K. Goto, J. Matsuo, T. Sugii, H. Minakata, I. Yamada, in *IEDM Tech Dig. 1996*, IEEE, (1996) pp. 435-438.

4. I. Yamada, "Development of Cluster Ion Beam Technology" in *IIT school book (2010)*, edited by J. Ziegler, (Ion Implantation Technology Co. Chester, MD, 2011) pp.13-1 - 13-62.
5. T. Aoki, N. Shimada, D. Takeuchi, J. Matsuo, Z. Insepov and I. Yamada, "The Molecular Dynamics Simulation of Boron Cluster Ion Implantation", IEICE Technical Report, 96, No.396, (1996) pp.49-54. (in Japanese).
6. Several papers in *18th International Conference on Ion Implantation Technology IIT 2010*, edited by J. Matsuo, M. Kase, T. Aoki and T. Seki, AIP Conference Proceedings 1321, (2010) and in *10th Workshop on Cluster Ion Beam Technology*, Kyoto, (2010).
7. C. Eggs, in *7th Workshop on Cluster Ion Beam Technology*, Tokyo, (2006) pp.46-51.
8. N. Toyoda and I. Yamada in *10th Workshop on Cluster Ion Beam Technology*, Kyoto, (2010) pp.75-78.
9. R&D Program & Technology Strategy Map (2007-2010), published by METI Japan. http://www.meti.go.jp/policy/economy/gijutsu_kakushin/kenkyu_kaihatu/str2010.html.
10. N. Toyoda, K. Nagato, H. Tani, Y. Sakane, M. Nakao, T. Hamaguchi, I. Yamada, J. Appl. Phys. 105, (2009) pp.07C127-1 - 07C127-2.
11. J. Weldon, R.MacCrimmon, S. Caliendo, Y. Shao, J. Hautala, B. Zide, M. Gwinn, in *5th Workshop on Cluster Ion Beam Technology*, Kyoto, (2004) pp.47-57.
12. I. Yamada, W. L. Brown, J. A. Northby and M. Sosnowski, Nucl. Instr. and Meth. B79, (1993) pp.223-226.
13. N. Toyoda, Kyoto University Ph.D Thesis (1999).
14. N. Toyoda, J. Matsuo, T. Aoki, S. Chiba, I. Yamada, D. B. Fenner and R. Tori, in *Ion Beam Synthesis and Processing of Advanced Materials*, edited by S. C. Moss, K. Heinig and D. Poker, (Mat. Res. Soc. Symp. Proc. 647, Pittsburgh, PA, 2001) pp.O5.1.1-O5.1.6.
15. S. Ninomiya, K. Ichiki, T. Seki, T. Aoki and J. Matsuo, Nucl. Instr. and Meth. in Phys. Res, B 256, (2007) pp.493-496.
16. S. Ninomiya, Y. Nakata, K. Ichiki, T. Seki, T. Aoki and J. Matsuo, in *10th Workshop on Cluster Ion Beam Technology*, Kyoto, (2010) pp.43-47.
17. K. Mochiji, M. Hashinokuchiy, K. Moritani and N. Toyoda, Mass Spectrom. 23, (2009) pp. 648–652.
18. T. Miyayama, N. Sanada, S. R. Bryan, J. S. Hammond and M. Suzuki, Surf. Interface Anal. 42, (2010) pp.1453–1457.
19. T. Albrektsson, P.I. Brånemark, H.A. Hansson, J. Lindström, Acta Orthop. Scand. 52, (1981) pp.155-70.
20. D. Buser, N. Broggini, M. Wieland, R.K. Schenk, A.J. Denzer, D.L. Cochran, B. Hoffmann, A. Lussi and S.G. Steinemann, J. Dent. Res. 83, (2004) pp.529-33.
21. T. Suzuki, N. Hori, W. Att, K. Kubo, F. Iwasa, T. Ueno, H. Maeda and T. Ogawa, Tissue Eng Pt A 15, (2009) pp.3679-3688.
22. H. Isogai, E. Toyoda, T. Senda, , K. Izunome, K. Kashima, N. Toyoda, and I. Yamada, in *16th International Conference on Ion Implantation Technology IIT 2006*, edited by K. Kirkby, R. Gwilliam, A. Smith and D. Chivers, AIP Conference Proceedings 866, (2006) pp.194-197.
23. A.G. Mikos, G. Sarakinos, M.D. Lyman, D.E. Ingber, J.P. Vacanti, R. Langer, Biotechnol. Bioeng. 42, (1993) pp.716-23.
24. A. Park, L.G. Cima, J. Biomed. Mater. Res. 31, (1996) pp.117-30.
25. T. Seki, T. Kaneko, D. Takeuchi, T. Aoki, J. Matsuo, Z. Insepov, I. Yamada, Nucl. Instr. and Meth. in Phys. Res. B. 121, (1997) pp.498-502.
26. G. Leseche, J. Ohan, S. Bouttier, T. Palombi, P. Bertrand, B. Andréassian, Ann. Vasc. Surg. 9 Suppl., (1995) pp.S15-23.
27. A. Yeager, A.D. Callow, ASAIO Trans. 34, (1988) pp.88-94.
28. F.R. Pu, R.L. Williams, T.K Markkula, J.A. Hunt, Biomaterials, 23, (2002) pp.2411-28.
29. D. Eddington, J. Puccinelli, D. Beebe, Sensors Actuators B-Chem. 114, (2006) pp.170-172.

Mater. Res. Soc. Symp. Proc. Vol. 1354 © 2011 Materials Research Society
DOI: 10.1557/opl.2011.1407

Nano-engineering with a focused helium ion beam

Diederik J. Maas[1], Emile W. van der Drift[2], Emile van Veldhoven[1], Jeroen Meessen[3], Maria Rudneva[2], and Paul F. A. Alkemade[2]

[1]TNO - van Leeuwenhoek Laboratory, Stieltjesweg 1, 2826 CK Delft, The Netherlands
[2]Kavli Institute of Nanoscience, Delft University of Technology, Lorentzweg 1, 2628 CJ Delft, The Netherlands
[3]ASML Netherlands B.V., de Run 6501, 5500 AH Veldhoven, The Netherlands

ABSTRACT

Although Helium Ion Microscopy (HIM) was introduced only a few years ago, many new application fields are budding. The connecting factor between these novel applications is the unique interaction of the primary helium ion beam with the sample material at and just below its surface. In particular, the HIM secondary electron (SE) signal stems from an area that is very well localized around the point of incidence of the primary beam. This makes the HIM well-suited for both high-resolution imaging as well as high resolution nanofabrication. Another advantage in nanofabrication is the low ion backscattering fraction, leading to a weak proximity effect. The lack of a quantitative materials analysis mode (like EDX in Scanning Electron Microscopy, SEM) and a relatively low beam current as compared to the SEM and the Gallium Focused Ion Beam are the present drawbacks of the HIM.

INTRODUCTION

The HIM scans sample surfaces with a sub-nanometer sized probe of fast helium ions[1]. Similar to the primary electron beam in SEM, at the sample the helium ions collide with surface and bulk atoms and thereby create secondary electrons (SEs). By recording the intensity of the SE signal while scanning the ion probe, the HIM creates an image of the sample surface with sub-nanometer resolving power[2, 3]. Figure 1 illustrates the main differences in interaction between SEM and HIM. The shorter wavelength of the heavier helium ion (as compared to electrons) enables one to focus to approximately the same spot size at a typically 5 times smaller numerical aperture. Hence, the depth-of-focus is much larger for HIM. The velocity of a 30 keV helium ion is comparable to that of a 4 eV electron. As a consequence, helium ions collide mainly with the valence electrons of the target atoms. These weak collisions generate hardly any X-rays or Auger electrons, and only a minor fraction of the helium is back-scattered, thereby restraining the analytical capabilities of the HIM[4]. The yellow areas in Figure 1 represent schematically the interaction volume of each primary beam within the sample. Livengood et al have simulated the shape and range of the interaction volume as a function of the primary particle energy for electrons, gallium and helium ions in silicon [5, 6]. As in Figure 1, Livengood's trajectory simulations show that primary electrons are (back-)scattered in a wider region than helium ions. Along their trajectory, the helium ions generate Secondary Electrons (iSE) with energy below 20 eV [7, 8], while primary electrons generate eSEs with energy up to a large fraction of the beam energy. In most materials, the mean free path of low energy SEs is of the order of a few nanometers [9]. Hence the escape depth of the iSEs is typically below 5 to 10 nm, whereas eSEs have a significantly larger escape depth. As a consequence, in HIM the interaction with the sample is more localized around the incident point of the scanning probe.

The red bars in Figure 1 qualitatively represent the sample area wherein detectable SEs are generated. Therefore HIM iSE images contain a relatively large amount of information at high spatial frequencies. This makes the HIM well-suited for both high-resolution imaging and also for high resolution nanofabrication [10-13].

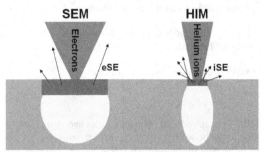

Figure 1 Schematic diagram illustrating the differences in the interaction of an electron beam or helium ion beam in a sample. For HIM the interaction with the sample is more localized around the incident point of the scanning probe. Therefore HIM iSE images contain a relatively large amount of information at high spatial frequencies.

To explore the nanofabrication capabilities with the focused helium ion beam, the HIM at the TNO Van Leeuwenhoek Laboratory (TNO-VLL)[14] is equipped with a pattern generator and a gas injection system (see Figure 2). For beam-induced processing, the tool is equipped with an OmniGIS (OmniProbe) gas injection system. This device houses reservoirs for three different reactive gases and has inputs for two carrier gases, to vary the flow and concentration of the active species. For all of the beam chemistry experiments reported here, deposition of platinum (bearing) deposits was generated from $MeC_pPt(IV)Me_3$ precursor (Colonial Metals). Etching was done with XeF_2 precursor from the same supplier. For maximum flexibility in defining writing strategies, the beam scan during deposition, etch and lithography can be controlled by the Elphy Plus (Raith GmbH) pattern generator.

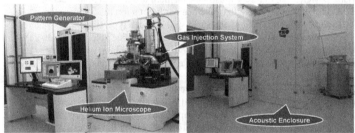

Figure 2 The HIM in TNO's van Leeuwenhoek Laboratory before and after the installation of the acoustic enclosure. The solid nitrogen cooling system allows the operator to work up to 12 hours at the highest resolution continuously. For nanofabrication purposes, the HIM is equipped with a Raith pattern generator and an OmniProbe gas injection system.

An ion spot that is small and bright, as well as beam pointing stability, are key in achieving (sub)nanometer resolution[15]. The HIM provides a beam current between 0.1 and 10 pA. Zeiss has developed an extremely low-vibration solid-nitrogen cryo-cooler for the ion source, which enables a continuous use of the HIM at its specifications [15]. The HIM at TNO-VLL is installed on a floating concrete floor that suppresses the transmission of low-frequency vibrations from the environment to the microscope. Additionally, TNO has developed an acoustic enclosure (see Figure 2) that suppresses room acoustic noise by at least 12 dB[16]. The amplitude of the unwanted vibrations in the HIM images is suppressed by at least a factor of 6.6 in the most important frequency band (200 to 3000 Hz). Even more important for nanofabrication, the enclosure ensures that every target pixel receives the same dose.

After describing an imaging application example addressing metrology, this paper will touch on the latest scanning helium ion beam lithography (SHIBL) results as well as nanostructures made with helium ion beam induced processes (HIBIP) like deposition and etching. SHIBL respectively HIBIP are discussed in depth in chapter 4 respectively 11 of the book *Nanofabrication, techniques and principles*[17]. The paper ends with a *Discussion and Conclusion* section that puts the HIM capabilities and the demonstrated application examples in a broader perspective.

EXPERIMENTS

In this section some application examples are discussed, illustrating the HIM's imaging and nanofabrication capabilities. For practical imaging applications, the optimum dose-per-area is determined by balancing the signal-to-noise ratio (S/N) in the recorded image with the sample integrity degradation due to beam-induced damage[18, 19]. The emphasis of imaging work at TNO-VLL[13] is on sensitive materials such as e.g. organic photo-voltaic materials and deep or extreme ultra-violet (DUV or EUV) resists. These samples have in common that they are difficult to image in a SEM due to their charging behavior and due to their sensitivity to both contamination and modification under electron beam irradiation. Sample charging is strongly reduced when imaging with ions. With a HIM high-resolution imaging is possible as sample charging is effectively cancelled with an electron flood gun. Besides the generation of iSEs, ion-sample interactions comprise helium implantation, neutralization, backscattering, and – although limited – some sputtering of target atoms. Roughly speaking, the iSEs are the most important part of the ion-sample interaction for both imaging and nanofabrication. For the HIM, several nanofabrication modes are available:

1. Direct Write (DW), i.e. sputter removal of target atoms by simple helium ion irradiation,
2. Helium Ion Beam Induced Deposition (HIBID), i.e. the creation of a deposit on the sample surface. The deposit consists of non-volatile precursor molecule fragments, which are created through local cracking by the ion beam,
3. Helium Ion Beam Induced Etching (HIBIE), i.e. the enhancement of removal of sample material by local cracking of adsorbed etchant molecules,
4. Scanning Helium Ion Beam Lithography (SHIBL), i.e. patterning of a resist layer on a substrate by inducing a chemical modification.

These four modes are illustrated in later sections of this paper. Nevertheless, we start with a challenging imaging application: metrology on EUV resist using SEM and HIM recordings of lines and spaces at 54 nm pitch.

Critical Dimension metrology on EUV resist with SEM and HIM

EUV chemically amplified resist (CAR) is known difficult-to-image in SEM due to charging, significant shrinkage during imaging, and beam-induced contamination. The Critical Dimension (CD) metrology accuracy and precision requirements for future nodes on the ITRS[20] roadmap impose significant imaging challenges for scanning probe microscopes[21-23]. Optical CD and CD-SEM tools are today's metrology workhorses and each have their own systematic offsets. These offsets become significant contributors to the total measurement uncertainty. The systematic errors of HIM have been investigated and compared to SEM by TNO and ASML. To this end, we acquired SEM and HIM images of a dense line pattern at 54-nm full-pitch written in EUV resist by ASML's EUV lithography demo tool at IMEC. Figure 3 shows examples of raw SEM and HIM images. The SEM image (left panel) and the HIM image in the central panel are recorded with 62.5 primary particles per pixel at a pixel size of 1 nm^2. Clearly, the HIM image provides better S/N. This improvement is due to the two-to-three times larger SE yield for helium ions than for electrons. The right HIM image in Figure 3 was recorded at 25 ions per pixel, corresponding to an areal dose of 10^{16} ions cm^{-2}. As a result, the S/N is similar to that of the SEM image. Remarkably, HIM images reveal a significantly lower resist line width and line-edge roughness than CD-SEM.

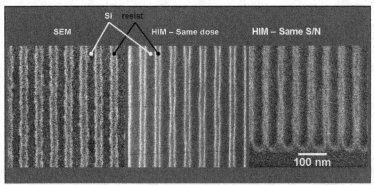

Figure 3 Images of dense lines and spaces at 54 nm full pitch. Left panel: Hitachi CD-SEM image as routinely used by ASML for metrology on EUV CAR resist. Center panel: HIM image recorded at the same pixel size and dose as the CD-SEM image. Due to the three times larger SE yield per primary particle, the S/N of this image is significantly higher. Right panel: HIM image recorded at same pixel size at three times lower dose. This image has a S/N comparable to that of the CD-SEM image. Still it shows significantly thinner and smoother resist lines, while the full pitch is the same for all images.

Figure 4 shows the change of the CD as a function of primary particle dose, derived from a sequence of images of the same area, recorded with SEM and HIM. At low doses, shrink of the resist occurs for both imaging modalities. At higher helium doses, resist swelling is observed. Images recorded at an angle, show that the swelling occurs both in height and width. A speculative explanation for the mechanism is that the scattering of the helium ions release hydrogen radicals from the Photo Acid Generator in the CAR. As a result, the internal forces (that determine the shape of the resist) change, leading to significant swelling. Extrapolation of the recorded CD to zero dose yields the best estimate of the line width in the resist. The difference in the extrapolated CD between SEM and HIM remains significant: 28 nm for SEM versus 20.5 nm for HIM. Presently, it is undecided which CD measurement tool is the more accurate.

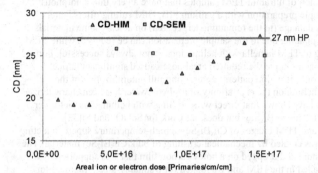

Figure 4 The measured CD from HIM and SEM images as a function of primary particle dose-per-area. The lithography process is tuned such that the first CD-SEM image matches with the design half pitch of 27 nm. Extrapolation to zero dose yields significantly different CD values for SEM and HIM, indicating the existence of systematic metrology errors that deserve further research and quantification, involving other metrology tools like OCD, and CD-AFM.

In conclusion, CD metrology with the HIM is at least as precise as CD-SEM. Yet, we expect additional benefits from HIM. Firstly, the different interaction mechanism allows for higher resolution imaging [24]. Secondly, the lower interaction volume causes smaller systematic errors in CD measurements, resulting in improved topology contrast. However, our conclusion and expectations are based on the results from *one* EUV and *one* DUV (not shown here) CAR resist, and hence they call for a study of a wider class of resists.

Transmission Electron Microscopy (TEM) lamella preparation with Direct Write milling

Gentle milling with the Helium ions can be used to create sub-10 nm wide cuts in a variety of materials. Several experiments on material modification at the nanometer scale demonstrate the high lateral confinement and hence local interaction of the helium ions with the sample: an 8 nm wide cut in Au, and a 5 nm square hole in a 100 nm thick Au film yield a 1:20 aspect ratio, and 5 nm wide graphene ribbons that are hundreds of nanometers long, and last but not least a very locally induced disorder in magnetic material [13, 25-29]. When compared to gallium-FIB milling, there are two significant differences. Firstly, the helium spot attributes, like probe size, shape and brightness, are smaller, better and higher, respectively, when compared to the spot properties of the best available gallium FIBs[30, 31]. Secondly, the low mass of the helium ion leads to a different energy- and momentum-transfer in ion-target atom collisions, as discussed in Chapter 12 in[17] and in [31, 32]. In practice, target atom sputtering is very effective at very high angles-of-incidence of the helium ion, yielding sputter rates per helium ion that even may exceed the yield per gallium ion. As a result, sputtering with helium can be used for the direct write of patterns with very shallow edges[26], and for the creation of TEM lamella with ultra-thin amorphous edges, especially when the incident helium ion beam direction is

37

normal to the target surface[33], see Figure 5. Most successful milling experiments are done on samples that initially are already thin, i.e. thinner than the stopping range of the 30 keV helium ions in the specific target material. Furthermore, the cutting of materials with nanometer precision enables the preparation of ultrathin TEM samples that have a very thin amorphous layer. Compared to TEM sample preparation with a gallium FIB, HIM cutting offers better precision combined with less damage (to the remaining target area) on at least some materials. As a result, artifacts in TEM imaging due to the sample preparation can be reduced, while TEM imaging is improved. Milling of TEM lamella with helium ions is not always successful. For example, when thinning a silicon sample the required areal dose caused significant lattice damage and yielded an electron diffraction pattern representing full amorphization of the exposed area. The impact of helium on the crystallinity of a silicon wafer has been characterized recently[5, 6]. Rudneva et al. have shown that direct write milling with helium ions yields very good TEM samples for Pt, Au and $Cu_xBi_2Se_3$, but does not work for $SrTiO_3$ and Si[33].

Figure 5 shows HIM and TEM images of $Cu_xBi_2Se_3$, a high-temperature superconducting material. The TEM lamella was created by mechanical crushing of bulk $Cu_xBi_2Se_3$ material. The fragments were mixed in a liquid and dropped on a holey carbon film that was supported by a golf grid. The sample was loaded in the HIM and we looked for a suitably-sized fragment for thinning by sputtering with helium. To unravel the charge transport mechanism, it is required to know the crystal structure of $Cu_xBi_2Se_3$. TEM images may resolve this structure, yet require thin samples (lamellas) that are oriented along the two symmetry planes. However, production of these lamellas is not easy, since $Cu_xBi_2Se_3$ is brittle along [001] planes and difficult to cleave in the [hk0] plane. We have thinned a flake of $Cu_xBi_2Se_3$ by locally irradiating the sample with helium ions. The HIM image in the left panel of Figure 5 shows the sputtering effect on the flake for four different ion doses. A slight incision is made at the lowest dose. At a slightly increased helium dose a partial cut through the flake occurs, while at the two highest doses a complete hole through the target was drilled. The structure and properties of $Cu_xBi_2Se_3$ as a function of the copper atoms are investigated in [34]. Our High Resolution TEM image shows a hexagonal pattern up to the HIM-made edge of the sample as can be seen in the right panel of Figure 5. The HRTEM image shows that the lamella is crystalline to the edge, indicating a very gentle milling in the HIM, possibly combined with some re-crystallization after or during the milling.

Alternative TEM lamella preparation methods like gallium ion milling and ultra-microtomy showed many surface deterioration related artifacts and thus HIM offers a very useful complementary addition to existing sample preparation possibilities.

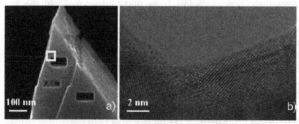

Figure 5 Example of Direct Write milling using helium ions in $Cu_xBi_2Se_3$ crystal by HIM and TEM images. The $Cu_xBi_2Se_3$ is crystalline up to the edge, indicating that little damage is inflicted during the sample thinning with the He ion beam.

Helium Ion Beam Induced Processing (HIBIP)

The process of beam-induced chemistry offers almost endless flexibility for both additive (by beam-induced precursor deposition) and subtractive (by beam-catalyzed etching) processing. In practice it is hard labor to find the right precursor and process conditions to reproducibly create a nanostructure. A recent review article by Utke et al. on electron and ion beam induced processes[35] makes evident, simply by its massive page count, that a large number of materials have been used to deposit nano and micro structures with a wide variety of conductive, insulating, magnetic, and photonic attributes. Many of these investigations aim to make prototype devices for applications in sensing, communication, storage or computing. Mask write and repair are presently the main industrial applications that require well-controlled deposition of free-form nanostructures[36]. Since the appearance of Utke's review in 2008, significant progress occurred in Helium Ion Beam Induced Processing (HIBIP)[12, 13, 37-39]. The following subsections show some of those advances in this rapidly expanding field.

Free-form pattern deposition with HIBID

The exact shape of the deposit depends on a number of factors like sample composition, precursor molecule supply, pattern scan settings and the intensity of the electron or ion beam [38, 40-42]. In the end, the precise control over the interaction of the helium ions with the target atoms enables the creation of nano-sculptures, see e.g. Figure 6. A HIBIP nano-reproduction of a 13th century drawing (left panel of Figure 6), representing the famous 8th century Chinese poet Li Bai, has been created as cover illustration for the PhD thesis of Ping Chen[40]. The image in the centre panel is recorded with the HIM. The right panel shows this HIM micrograph with inverted grey values. The qualities of this 21st century nano-copy come remarkably close to those of the 13th century original, including the variations in grey levels. This gradation from black-to-white was obtained by optimizing the dose and beam scan strategy (pixel dwell time and number of repeats). A more systematic study on the fundamentals of nano- pillar growth from deposition processes using both gallium and helium ions is given in [40].

Figure 6 A HIBIP deposit (center panel: HIM image , right panel: inverted grey values) representing a 100.000 times reduced copy of a 13th century painting by Liang Kai (left panel) representing 8th century poet Li Bai.

Figure 7 shows a reproduction of the TNO company logo at the nanometer scale from the Pt HIBID process. When comparing the deposit with the original, a number of observations can be made, based on the distortion of the deposit shape. Firstly, the smallest features in the deposit are the slogan letters, which are each approximately 20 nm wide. The corresponding features in the original measure 12 nm on this scale. The hole in the *e* in *life* is just closed due to the overspray, while the *a* and *os* in *innovation* are still open. It can be concluded that the deposit is systematically broadened on the pattern edges by approximately 4-5 nm. The capitals **T** and **N** also illustrate the lower size limit of HIBIP deposits. In the original, these capitals have both square and rounded corners. In the deposit, all corners have a finite roundness with a radius of curvature of circa 10-12 nm, remarkably close to half the minimum line width. Such systematic broadening indicates that the deposit growth rate was limited by the amount of available precursor molecules, and not by the helium ion beam current. Secondly, the lateral extension of the deposit is of a magnitude equal to that of the mean free path of the iSEs in matter[9]. This suggests that a significant fraction of the precursor molecules is dissociated by the iSEs, and not directly by the helium ions. Monte Carlo simulations of the pillar growth process support this hypothesis [43, 44]. This also means that this nano-patterning method has faced its intrinsic resolution limit. Thirdly, the gap between the letters is as small as 6 nm. Together with the good separation of individual letters and the absence of a halo around the deposit, this demonstrates the very low proximity effects that come with the interaction of the helium ions. Once characterized, the systematic distortion due to the processing artifacts can be accounted for by pre-correction of the scan strategy and pattern.

Figure 7 Bottom panel: the TNO company logo at the nm scale. The small letters are on average 50 nm wide.

Patterning the TaN absorber on EUV masks with HIBIE

The Semiconductor industry deploys advanced mask design technology to push the limit of 193 nm as well as to make EUV lithography ready for high-volume manufacturing[45]. At both actinic wavelengths, this raises technology challenges for mask manufacturing. DUV lithography comes with several phase shift embodiments, which all have in common that the mask must be patterned with sub-wavelength assist features. EUV lithography comes with reflective masks which involves the development of novel patterning technology.

When targeting mask write and/or repair for future lithography nodes, it is of importance to have the capability to add or remove the photon absorbing material on the reticles. The previous section illustrated some of the deposition capabilities with the HIM. TaN is a commonly used absorber for EUV photons as used on standard EUV reticles. The precise and local removal of TaN and its native oxide TaNO on a EUV test reticle piece by Direct Write and HIBIE is shown in Figure 8. The efficacy of the XeF_2 precursor gas in removing the TaN is investigated by applying the same dose and scan strategy with and without the precursor gas flowing. For each case, a pattern of six single pixel line scans at 25 nm pitch was exposed 5500 times using a pixel dwell time of 1 microsecond, a pixel size of 1 nm, at an ion beam current of 0.6 pA, corresponding to a total areal ion dose of 300 mC cm^{-2} or 2 x 10^{18} ions cm^{-2}. With DW, some patterning of the TaNO surface is visible (left panel of Figure 8). The etched line width is 5 nm at a pitch of 25 nm. With HIBIE, the TaNO top layer is completely removed at the exposed pattern (right panel of Figure 8). The line width has broadened to 12 nm. A proper comparison of the TaN volume removal rate demands for height measurements of the trenches, e.g. by AFM, or a using a sensitive mass sensor as demonstrated in [46]. From the present experiment, however, no valid etch rate can be determined, since XeF$_2$ is known to etch TaN spontaneously. Hence, once the top TaNO layer is pierced, TaN removal occurs without the involvement of helium ions, thus enhancing the removed volume. To determine the HIBIE removal rate properly, another experiment is needed in which the exposed and etched surface is pacified instantly, e.g. by properly mixing the XeF$_2$ gas with an oxidizing carrier gas like O$_2$ or H$_2$O[47]. The present experiment shows that HIBIE is capable of patterning EUV reticles with 12 nm lines at a pitch, which will meet the lateral resolution requirement of EUV reticles until perhaps even the 3 nm node, which is presently not yet at the ITRS[20].

Figure 8 HIM images of the TaNO capping layer that is regularly applied as an Extreme Ultra Violet (EUV) photon absorber in EUV lithography. The initially uniform TaN(O) layer is exposed to the same line dose of helium ions without (left panel) and with (right panel) an etchant precursor molecule (XeF$_2$) in the sample chamber. Clearly, the mask patterning speed is strongly enhanced by the XeF$_2$. The written pattern shows 250 nm long and 12 nm wide lines on a 25 nm pitch, demonstrating low proximity effects.

<u>Scanning helium ion beam lithography</u>

A useful review including all pros and cons when comparing ion- with electron- beam lithography is given by Melngailis[48]. A recent breakthrough towards sub-10 nm patterning with ions came with two recent He$^+$ litho studies on HSQ[39, 40]. Chapter 11 in [17] provides an in-depth comparison of EBL with SHIBL. The present paper brings the latest result on scanning helium ion beam lithography (SHIBL) using negative tone hydrogen sil-sesquioxane (HSQ) resist. Figure 9 shows well-developed 5 nm lines on a pitch of 10 nm after SHIBL exposure in a 5 nm thick resist layer. The resolution performance is more or less equivalent to the best electron beam lithography (EBL) achievement[49, 50]. Moreover, SHIBL clearly features superior reduction of proximity effects as compared to Electron Beam Lithography (EBL): the isolated line has the same width as the lines in the dense pattern, while all lines were exposed at the same dose. The low proximity effects in SHIBL can be explained from the scattering profile of the ions in the substrate: the trajectories of helium ions are much more forward directed, as compared to electron trajectories. In addition, the almost negligible backscattering of helium greatly reduces proximity effects from adjacent pixel exposures. Together, these helium scattering attributes allow for the breakthrough in achievable pattern density shown in Figure 9 where in EBL electron backscattering tends to be a show-stopper[49]. For the actual exposure there is another benefit: resists have a one-to-two orders higher sensitivity for helium as compared to electrons. This reduces the damage to the target substrate which occurs from the stopping collisions of helium ions with the target atoms. Also the throughput of the exposure is increased, which partially makes up for the lower available ion beam current. The increased resist sensitivity can be explained from two attributes of the interaction of helium ions within the resist, as compared to electron beam lithography[51]. An important aspect of particle beam lithography is the generation of secondary electrons (SEs), which in both SHIBL and EBL account for the actual exposure reaction. Firstly, the electronic stopping in a thin (below 100 nm) resist layer of 30 keV Helium ions is about 90 times stronger, as compared to 30 keV electrons[52]. It is assumed that for both primary particles a similar fraction of the dissipated energy is converted into secondary electrons. Secondly, several studies show that for He^+ the SE energy spectrum is shifted to lower energies as compared to SE from electron exposure[3, 8]. Depending on the resist-specific activation energy, iSEs have a 2-3 times lower cross-section for activation of the chemical reaction as compared to eSEs, when averaged over their full energy spectrum[51]. These effects combined roughly explain the two orders of magnitude higher sensitivity of resists for helium ions: the ion stopping power is 90 times higher, but eSEs are 2-3 times more effective iSEs: $90/2,5 \cong 35$. The remaining mismatch between this hypothesis and the observed sensitivities is attributed to some quantitatively unknown factors, e.g. the energy dependence of the cross section for resist bond activation and/or the variation in the number of generated SEs during the stopping of the primary particles.

Yet, the most important feature of SHIBL is its capability to write fine and dense patterns, due to the strong forward scattering and negligible backward scattering of the helium ions as compared to electrons.

Figure 9 SEM image of 5 nm wide lines at a pitch of 10 nm in a 5 nm thick HSQ resist made by 30 keV helium ion beam lithography. The line dose of 0.12 nC cm^{-1} is applied with a beam step size of 1 nm. Note the absence of proximity effects: the isolated segment of the long line has the same width as the nested fragment.

DISCUSSION and CONCLUSIONS

The HIM is a novel high-resolution imaging and nanofabrication tool, with some rather remarkable capabilities. The unique interaction of the helium ions with the sample can be applied for imaging, patterning and modification of the target. This paper has presented several examples of nano-engineering with the high-brightness focused helium ion beam of the HIM.

For imaging, HIM unlocks a new application space especially for imaging sensitive (charging) surfaces with (sub)nm resolution. Like with many scanning microscopes that have a nanometer resolving power (like e.g. STEM, SEM and AFM), while generating desired effects, the interaction of the primary beam with the sample can cause damage to the target. Fortunately, also for the HIM any unwanted sample damage often occurs at much higher doses than are required for imaging, lithography and beam chemistry.

The helium ion microscope shows a strong ability to produce fine and dense features reproducibly. The low backscattering of the primary particles enables ultra-fine and -dense patterning using beam induced processes, and in resist. Further optimization of SHIBL and HIBIP involves many experimental factors relating to the HIM settings, sample attributes, (post-) processing conditions and the final inspection method. All these factors are to be considered, evaluated and controlled to reproducibly deliver nano-products and processes with desired properties, such as e.g. critical dimension, yield-per-ion, resistivity and compositional purity. Experiments guided by a Design-Of-Experiment approach assist in the systematic identification of significant factors, which in turn provide valuable hints for improving the instrumentation, recipes, and metrology involved. Last but not least, the resolution of nano-patterning with SHIBL and HIBIP is not limited by the size of the ion probe but by the intrinsic length scale of the interactions of the ions and the generated secondary electrons with the resist, the precursor molecules and the substrate. In fact, a major limitation at this point lies with the metrology and analysis of the results. Profile measurements by AFM would be a logical next step to get the most accurate measurements. Making accurate measurements at these length scales will provide interesting challenges for some time to come.

Now that the HIM is commercially available and has turned into a practical instrument that yields results reproducibly, the largest challenge for scientists, engineers and marketers is in identifying useful applications that could not be achieved in the past, when the HIM did not exist yet. A similar challenge will be raised once the Neon Ion Microscope, which is currently in development[15], is introduced in the market. It is to be expected that combination of a bright focused probe and the specifics of the interaction of Neon ions with samples will once again open up new unique applications for nano-imaging and -fabrication.

ACKNOWLEDGMENTS

All of this work would not have been possible without a significant investment funded by the Dutch NanoNed program. ASML has financially supported the CD metrology benchmark of SEM with HIM. The HREM application is supported by NIMIC. Zeiss has financially and scientifically (Larry Scipioni, Colin Sanford, Lewis Stern and David Ferranti) supported the HIBID experiments.

At the date this paper was written, URLs or links referenced herein were deemed to be useful supplementary material to this paper. Neither the author nor the Materials Research Society warrants or assumes liability for the content or availability of URLs referenced in this paper.

REFERENCES

1. Morgan, J., et al., *An Introduction to the Helium Ion Microscope.* Microscopy Today, 2006. **14**(4): p. 24-30.
2. Vladár, A.E., M.T. Postek, and B. Ming, *On the Sub-Nanometer Resolution of Scanning Electron and Helium Ion Microscopes.* Microscopy Today, 2009. **17**: p. 6.
3. Scipioni, L., et al., *Understanding imaging modes in the helium ion microscope* J. Vac. Sci. Technol. B, 2009. **27**.
4. Joy, D.C. and B.J. Griffin, *Is Microanalysis Possible in the Helium Ion Microscope?* Microscopy and Microanalysis, 2011. **17**(4): p. 643-649.
5. Livengood, R., et al., *Subsurface damage from helium ions as a function of dose, beam energy, and dose rate.* Journal of Vacuum Science & Technology B: Microelectronics and Nanometer Structures, 2009. **27**(6): p. 3244-3249.
6. Tan, S., et al., *Gas field ion source and liquid metal ion source charged particle material interaction study for semiconductor nanomachining applications.* Journal of Vacuum Science & Technology B: Microelectronics and Nanometer Structures, 2010. **28**(6): p. C6F15-C6F21.
7. Ramachandra, R., B. Griffin, and D. Joy, *A model of secondary electron imaging in the helium ion scanning microscope.* Ultramicroscopy, 2009. **109**(6): p. 748-757.
8. Petrov, Y. and O. Vyvenko, *Secondary electron emission spectra and energy selective imaging in helium ion microscope*, in *SPIE Scanning Microscopies 2011: Advanced Microscopy Technologies for Defense, Homeland Security, Forensic, Life, Environmental, and Industrial Sciences* 2011, SPIE.
9. Seah, M.P. and W.A. Dench, *Quantitative electron spectroscopy of surfaces: A standard data base for electron inelastic mean free paths in solids.* Surface Interface Analysis, 1979. **1**: p. 1.
10. Postek, M.T., A.E. Vladár, and B. Ming, *Recent progress in understanding the imaging and metrology using the helium ion microscope*, in *Proc. SPIE*, M.T. Postek, et al., Editors. 2009, SPIE: Monterey, CA, USA. p. 737808-10.
11. Postek, M.T., et al., *Review of current progress in nanometrology with the helium ion microscope.* Meas. Sci. Technol., 2011. **22**(2).
12. Alkemade, P.F.A., et al., *Model for nanopillar growth by focused helium ion-beam-induced deposition.* Journal of Vacuum Science & Technology B: Microelectronics and Nanometer Structures, 2010. **28**(6): p. C6F22-C6F25.
13. Maas, D.J., et al., *Nanofabrication with a helium ion microscope.* SPIE Metrology, Inspection, and Process Control for Microlithography XXIV 2010. **7638**: p. 763814.
14. TNO-VLL. Available from: http://www.vanleeuwenhoeklab.com/.
15. Hill, R. and F.H.M. Rahman, *Advances in helium ion microscopy.* Nucl. Instr. and Meth.A, 2010.
16. van Beek, P.J.G., et al., *Acoustic immunity improvement for the Helium Ion Microscope - Private communication.*
17. Stepanova, M. and S. Dew, eds. *Nanofabrication: Techniques and Principles.* 1st Edition ed. 2011, Springer: Wien. 4223.
18. Orloff, J., L.W. Swanson, and M. Utlaut, *Fundamental limits to imaging resolution for focused ion beams* J. Vac. Sci. Technol. B, 1996. **14**: p. 5.
19. Castaldo, V., et al., *On the influence of the sputtering in determining the resolution of a scanning ion microscope.* J. Vac. Sci. Technol. B, 2009. **27**: p. 3196.

20.   IRC. Available from: http://www.itrs.net/.
21.   Postek, M.T., A.E. Vladár, and B. Ming, *Breaking the resolution barrier: understanding the science of helium ion beam microscopy.*, in *Frontiers of Characterization and Metrology for Nanoelectronics*, D.G. Seiler, et al., Editors. 2009, AIP. p. 249–60.
22.   Postek, M.T. and A.E. Vladár, *Helium ion microscopy and its application to nanotechnology and nanometrology.* Scanning, 2008. **30**(6): p. 457-462.
23.   Jepson, M., et al., *Resolution Limits of Secondary Electron Dopant Contrast in Helium Ion and Scanning Electron Microscopy.* Microscopy and Microanalysis, 2011. **17**(04): p. 637-642.
24.   Ohya, K., et al., *Comparison of secondary electron emission in helium ion microscope with gallium ion and electron microscopes.* Nuclear Instruments and Methods in Physics Research Section B: Beam Interactions with Materials and Atoms, 2009. **267**(4): p. 584-589.
25.   Yang, J., et al., *Rapid and precise scanning helium ion microscope milling of solid-state nanopores for biomolecule detection.* Nanotechnology, 2011. **22**: p. 285310.
26.   Scipioni, L., et al., *Fabrication and initial characterization of ultrahigh aspect ratio vias in gold using the helium ion microscope* J. Vac. Sci. Technol. B 2010. **28**: p. C6P18.
27.   Bell, D.C., et al., *Precision cutting and patterning of graphene with helium ions.* Nanotechnology, 2009. **20**.
28.   Pickard, D. and L. Scipioni, *Graphene Nano-Ribbon Patterning in the Orion Plus".* Zeiss application note, 2009.
29.   Franken, J.H., et al., *Precise control of domain wall injection and pinning using helium and gallium focused ion beams.* Vol. 109. 2011: AIP. 07D504.
30.   Orloff, J., L. Swanson, and M. Utlaut, *High Resolution Focused Ion Beams: FIB and Its Applications.* 2003, New York: Springer Press. 316.
31.   Castaldo, V., *High resolution scanning ion microscopy.* 2011, Delft University of Technology.
32.   Castaldo, V., et al., *Angular Dependence of the Ion-Induced Secondary Electron Emission for He⁺ and Ga⁺ Beams.* Microscopy and Microanalysis, 2011. **17**(04): p. 624-636.
33.   Rudneva, M., et al., *HIM lamella preparation*, in *ICM17*. 2010: Rio de Janeiro, Brazil.
34.   Hor, Y.S., et al., *Superconductivity in $Cu_xBi_2Se_3$ and its Implications for Pairing in the Undoped Topological Insulator.* Physical Review Letters, 2010. **104**(5): p. 057001.
35.   Utke, I., P. Hoffmann, and J. Melngailis, *Gas-assisted focused electron beam and ion beam processing and fabrication.* Journal of Vacuum Science & Technology B: Microelectronics and Nanometer Structures, 2008. **26**(4): p. 1197-1276.
36.   Edinger, K., et al., *Electron-beam-based photomask repair.* Vol. 22. 2004: AVS. 2902-2906.
37.   Sanford, C., et al., *Beam induced deposition of platinum using a helium ion microscope.* Journal of Vacuum Science and Technology B: Microelectronics and Nanometer Structures, 2009. **27**(6): p. 8.
38.   Scipioni, L., et al., *A Design-of-Experiments Approach to Characterizing Beam-Induced Deposition in the Helium Ion Microscope.* Microscopy Today, 2011. **5**: p. 7.
39.   Boden, S.A., et al., *Focused helium ion beam milling and deposition.* Microelectronic Engineering, 2011. **88**(8): p. 2452.

40. Chen, P., *Three-dimensional Nanostructures Fabricated by Ion-Beam-Induced Deposition*, in *Kavli Institute for Nanoscience*. 2010, Delft University of Technology: Delft.

41. Plank, H., et al., *The influence of beam defocus on volume growth rates for electron beam induced platinum deposition*. Nanotechnology, 2008. **19**: p. 485302.

42. Ebm, C., et al., *Modeling of precursor coverage in ion-beam induced etching and verification with experiments using XeF_2 on SiO_2*. Journal of Vacuum Science & Technology B: Microelectronics and Nanometer Structures, 2010. **28**(5): p. 946-951.

43. Chen, P., et al., *Nanopillar growth by focused helium ion-beam-induced deposition*. Nanotechnology, 2010. **21**: p. 455302.

44. Smith, D.A., D.C. Joy, and P.D. Rack, *Monte Carlo simulation of focused helium ion beam induced deposition* Nanotechnology, 2010. **21** p. 175302.

45. Rizvi, S., ed. *Handbook of photomask manufacturing technology*. 2005, Taylor & Francis: Boca Raton.

46. Friedli, V., et al., *Mass sensor for in situ monitoring of focused ion and electron beam induced processes*. Applied Physics Letters, 2007. **90**(5): p. 053106-053106-3.

47. Lassiter, M.G., T. Liang, and P.D. Rack, *Inhibiting Spontaneous Etching of Nanoscale Electron Beam Induced Etching Features: Solutions for Nanoscale Repair of Extreme Ultraviolet Lithography Masks*. Journal of Vacuum Science & Technology B, 2008. **26**: p. 3.

48. Melngailis, J., *Focused Ion Beam Lithography (review article)*. Nuclear Instrum. Methods in Phys. Res., 1993. **B80**: p. 1271

49. Yang, J.K.W., et al., *Understanding of hydrogen silsesquioxane electron resist for sub-5-nm-half-pitch lithography* J. Vac. Sci. Technol. B, 2009. **27**: p. 6.

50. Sidorkin, V., *Resist and Exposure Processes for Sub-10-nm Electron and Ion Beam Lithography*. 2010, Delft University of Technology.

51. van Langen-Suurling, A., et al., *Nanolithography with a scanning sub-nanometer helium ion beam*. Submitted to J. Vac. Sci. Technol. B.

52. Berger, M.J., et al., *ESTAR, PSTAR, and ASTAR: Computer Programs for Calculating Stopping-Power and Range Tables for Electrons, Protons, and Helium Ions (version 1.2.3)*. 2005, National Institute of Standards and Technology: Gaithersburg, MD. http://physics.nist.gov/Star

Mater. Res. Soc. Symp. Proc. Vol. 1354 © 2011 Materials Research Society
DOI: 10.1557/opl.2011.1210

Multi-Ion Beam Lithography and Processing Studies

Bill R. Appleton[1], Sefaattin Tongay[1], Maxime Lemaitre[1], Brent Gila[1], David Hays[1], Andrew Scheuermann[1], and Joel Fridmann[2]

[1]NIMET Nanoscale Research Facility, University of Florida, Gainesville, FL. 32611
[2]Raith USA, Inc., Ronkonkoma, NY, 11779

ABSTRACT.

The University of Florida (UF) have recently collaborated with Raith Inc. to modify Raith's ion beam lithography, nanofabrication and engineering (ionLiNE) station that utilizes only Ga ions, into a multi-ion beam system (MionLiNE) by adding the capabilities to use liquid metal alloy sources (LMAIS) to access a variety of ions and an EXB filter for mass separation. The MionLiNE modifications discussed below provide a wide range of spatial and temporal precision that can be used to investigate ion solid interactions under extended boundary conditions, as well as for ion lithography and nanofabrication. Here we demonstrate the ion beam lithographic capabilities of the MionLiNE for fabricating patterned arrays of Au and Si nanocrystals, with nanoscale dimensions, in SiO2 substrates, by direct implantation; and show that the same direct-write/maskless-implantation features can be used for in situ fabrication of nanoelectronic devices. Additionally, the spatial and temporal capabilities of the MionLiNE are used to explore the effects of dose rate on the long-standing surface morphological transformation that occurs in ion bombarded Ge.

INTRODUCTION.

The MionLiNE system shown in Figure 1 combines a precision scanned focused ion beam with a laser interferometer sample handling stage and lithographic patterning software for an integrated system that delivers mass-separated, multi-ion beams for processing. The system has a vacuum load lock for sample handling and an internal optical microscope to facilitate initial set up and navigation

The actual ion column and its representative schematic are shown in Figure 2. The top flange holds the LMAIS that can be isolated from the rest of the system and LMAIS sources interchanged. The EXB Wien filter consists of a permanent magnet and a variable electric field. The permanent magnet is located outside the column/vacuum and is easily interchangeable should the desired ion

Figure 1. MionLiNE system.

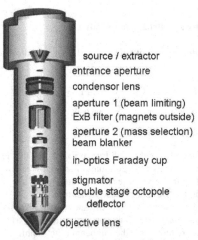

source / extractor
entrance aperture
condensor lens
aperture 1 (beam limiting)
ExB filter (magnets outside)
aperture 2 (mass selection)
beam blanker

in-optics Faraday cup

stigmator
double stage octopole
deflector

objective lens

Figure 2. Schematic representation of the ion beam column optics, and an actual photograph of an installed system at right .

selection require a different permanent field. The mass separation capability of the current filter is m/dm> 35, and the acceleration voltage can be adjusted from V= 15 – 40 kV. The LMAIS reported on in this paper is AuSi, and as the mass spectrum in Figure 3 shows it is possible to resolve singly and doubly charged ions as well as ion clusters of both Si and Au. Ion

energies are E=qV where q is the ion charge, so doubly charged ions allows doubling the implantation energy; and ion clusters with n atoms can provide reduced energies E= qV/n as the atoms separate on entering the solid. Beam dimensions can be adjusted with limiting apertures ranging from 5μm to 1000μm in diameter producing currents from picoamps to nanoamps depending on on the operating conditions. The beam current stability of the system is better than 1%/hour.

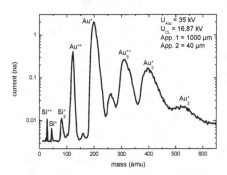

Figure 3. Mass spectrum from the AuSi source using listed parameters.

Sample handling is facilitated by the vacuum load lock, and an internal optical microscope for coarse positioning. Precision manipulation is accomplished with the Raith laser interferometer stage shown in Figure 4, this is driven by the Raith Motor controller, which is capable of resolving the stage movement to 1nm. A 20 MHz 16 bit pattern processor controls beam deflection. These make the MionLiNE system a lithography

Figure 4. MionLiNE sample holder positioned with a laser interferometer stage.

tool capable of direct write focused ion beam patterning at nanometer resolution and accuracy over large areas of 100 x 100 mm². Consequently, blind writes can be performed on sample areas up to 100 mm away with stitching and overlay accuracies ~ 30nm and 1 nm step resolution.

This sample stage, pattern generator, and software also provides for a fixed-beam moving stage (FMBS) feature that is illustrated in Figure 5. In this mode scanning is turned off and the ion beam is held fixed; and the laser stage is programmed to move the sample beneath the beam. This makes it possible to create virtually any pattern for lithography or sputter sculpting over large areas without the stitching errors that arise with conventional systems. The pattern in Figure 5 was written with 30keV Au in a GaAs crystal by starting at the outer diameter, spiraling into the center and then spiraling back out to the starting point parallel to the first write. The line-width was 30nm, the line is ~ 100 mm long and took 45 minutes to write.

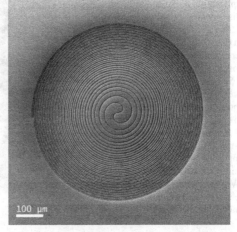

Figure 5. Spiral demonstration pattern written in the FBMS mode with the MionLiNE.

EXPERIMENTAL RESULTS AND DISCUSSION: AU AND SI NANOCRYSTALLINE ARRAYS.

Extensive research has been done on patterned metal and silicon nanocrystal arrays using ion implantation and other techniques in a variety of host materials [1-8] because of potential applications in optoelectronics, hybrid electronics, biotechnology and medical diagnostics, magnetic recording and information storage, on-chip photonics, catalytic and mechanical enhancements, and other uses [9-17]. The advantages of ion implantation are well known (implanting virtually any species of ion – even isotopes- into virtually any substrate; controlled concentrations and depth profiles; potential metastable phase formation and/or nonequilibrium concentrations; etc.). However, if the goal is to fabricate controlled nanoscale structures then it becomes necessary to control the species, size, concentration, and spatial patterning of the ion beam and implantation process. This can be accomplished using conventional lithography techniques combined with broad-area implantation but this can require numerous integrated processing steps (e.g. masking, etching, liftoff, deposition, implantation, etc.). Such processes have been successfully combined with ion implantation for spatial patterning with micron-dimensions to form luminescent arrays [6, 7, 18, 19], but this can require special modifications of conventional processing steps such as applying metal instead of resist masks to mitigate the effects of high-fluence ion bombardment.

One application of the MionLiNE system has been to fabricate nanocrystalline arrays by direct-write, maskless implantation of Au and Si beams [20]. This is demonstrated in the TEM micrograph shown in Figure 6 where 48 nm diameter ion beams of 30 keV Au were implanted into 40 nm thick SiO_2 films in a 16x16 array of spots separated by 500 nm. This was easily and quickly accomplished without any additional processing steps and other beam sizes, ion species, and patterns could have been accomplished with the MionLiNE system. For example, since the self-supporting SiO_2 sample had multiple windows it was possible to switch to 15 keV Si beams in situ without removing the sample and implant similar Si nanocrystalline arrays.

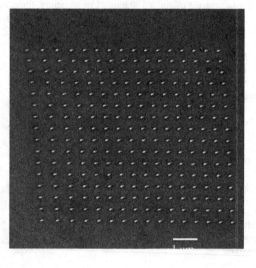

Figure 6. TEM micrograph of a nanocrystalline array of 48 nm diameter Au clusters implanted by direct-write, maskless implantation using the MionLiNE. (See text for details).

A variety of SiO2 films were implanted and some annealed to 950 C in vacuum for 30 minutes and others by rapid thermal annealing in Ar for 10 seconds. Nanocrystals of Au and Si were formed in all cases examined. TEM analysis of one of the implanted 48 nm diameter Au spots shown in Figure 6 was performed and typical results are reproduced in Figure 7. Microdiffraction and line spacing measurements showed that the nanocrystals, ranging in size from 2 to 8 nm, were crystalline.

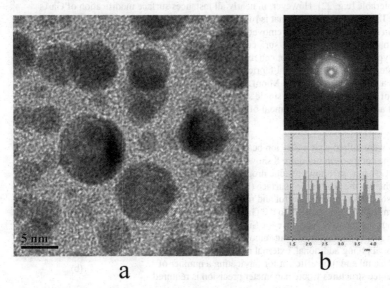

a   b

Figure 7. TEM analysis showing Au nanocrystals (a) formed by direct implantation into 40 nm thick SiO_2 films (see text). Micrographs and line scans (b) were made to verify that the Au clusters were crystalline and to determine orientations.

The 15 keV Si implants were performed in the same manner, in similar substrates, with similar fluences implanted into separate windows, with ion beam currents of 2.7 pA, and were annealed under the same conditions as Au; but in 10 x 10 arrays separated by 500 nm. As with Au the implanted spots were all consistent and the implanted arrays regularly arranged as programmed into the system software. Examination by TEM of one of the Si implants at incident fluences of 5 x 10^{16} Si ions/cm^2 showed that Si nanoclusters were formed [20].

EXPERIMENTAL RESULTS AND DISCUSSION: FABRICATING NANOELECTRONIC DEVICE FEATURES.

The capabilities of the MionLiNE for fabricating features in GaAs prototype devices has also been explored [21]. Gallium Arsenide is widely used for semiconductor device applications and is often processed using conventional FIBs with Ga sources since the Ga introduced in the process is tolerable [e.g. 22]. However, in nearly all instances surface modification of GaAs using a Ga FIB results in a surface that is rough and has a high density of Ga droplets that most be removed by some post-processing that can further affect the surface. Similarly, Si is a common n-type dopant for GaAs that can require numerous processing steps when a conventional fabrication approach is taken. So as a further exercise of the MionLiNE system the availability of Au and Si beams in a single system was investigated for potential advantages for development of prototype GaAs devices.

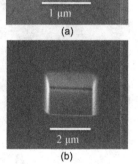

(a)

GaAs boxes were milled with a Ga+ ion beam and with an Au+ ion beam. The comparison in Figure 8 shows that the GaAs surface milled with Ga+ ions (a) shows residue droplets of gallium which is typical, whereas the Au+ milled surface (b) is clean and free of droplets. Also, the ion column control and reproducibility made it possible to mill alignment marks into the GaAs surface with the Au+ ion beam, switch to Si, and have the Si ion beam ready for implantation in a few minutes without removing the sample from the system or applying additional, external masking steps. This has proved very useful and time efficient for fabricating a number of prototype device structures where nanometer precision is required.

(b)

Figure 8. SEM images comparing GaAs boxes milled with: (a) a 6pA current Ga+ ion beam (1 μm square box tilted 45 degrees); (b) a 30pA current of Au+ beam (2μm square box tilted 35 degrees). Note Ga droplets visible at the bottom of (a).

EXPERIMENTAL RESULTS AND DISCUSSION: EVIDENCE FOR DOSE RATE DAMAGE EFFECTS IN GE.

The precise control of MionLiNE's time and space parameters and its multi-ion capabilities offer some interesting opportunities for studying ion solid interactions. In preparation for this Symposium we decided to re-visit the long-standing ion damage effect in Ge that results in a drastic transformation of the surface morphology.

As a refresher, this effect was first observed in 1982 [23,24], and has been widely studied since [25 - 31]; and while many of the details associated with the effect have been quantified and models proposed in these studies, a full understanding of how this effect evolves and why the particular surface morphology results are still worthy of investigation.

The evolution of this effect is shown in Figure 9 for a Ge(100) single crystal bombarded with 60 keV Au ions for several fluences at room temperature. These images were taken with an FEI Nova nanoSEM 430, at 5kV, and all at 100,000 x magnification.

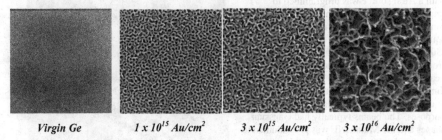

Virgin Ge   *1 x 10^{15} Au/cm^2*   *3 x 10^{15} Au/cm^2*   *3 x 10^{16} Au/cm^2*

Figure 9. SEM images of a Ge (100) crystal bombarded with a 250 nm diameter, 60 keV Au beam, scanned to achieve uniform fluences as shown.

Previous studies have shown that this effect:

➤ Varies with bombarding ion mass (species), energy, dose/fluence, and sample temperature during bombardment

➤ Is pronounced for ions ~ Ge and heavier, and ion clusters; but not for light ions

➤ Has been observed for Ge, As, Kr, In, Sn, Sb, Xe, W, Au, Tl, Pb, and Bi; and so is not likely any implanted material or alloying effect

➤ Is suppressed at low temperatures and above 325 C; but is pronounced from ~ RT to ~ 250 C

➤ Only occurs in the amorphous phase of Ge and corresponding temperature range

➤ Is independent of the angle of implantation, and a free surface is not required for formation [28]

➤ Shows surface density decreases up to ~70%, and swelling continues to grow with increasing fluence

➤ Has shown ~ 30-70% loss of implanted dopants above that expected from sputtering [23, 25, 26]

This range of phenomena that contribute to the effect and their interdependences require considerable control of experimental conditions before valid interpretations of the resulting measurements can be made.

To investigate this effect under varying fluence and time delivery conditions, the MionLine with a AuSi source was programmed to implant nine 10μm x 10μm squares spaced 20 μm apart like those shown in Figure 10. Each square was implanted by a 60 keV Au beam (30kV Au^{++}) with a different fluence ranging from $1x10^{13}$ to $3x10^{16}$ Au ions/cm^2. The Au beam was expanded to 250 nm diameter and its dimensions determined from 20-80% line scans over a Sn "ball" standard used for such purposes. The beams typically had Gaussian shapes and the MionLiNE was programmed to overlap and position these beams in each 10μm x 10μm area in such a way as to produce a uniform fluence in each square. In this arrangement the ion beam is incident on a spot within the square – pixel – and implants that spot with a programmed dose determined by a dwell time t_d ; the beam is then switched to an adjacent pixel where the same implanted dose is repeated. The pixel spacing is programmed to overlap the previous spot(s) in such a way as

Figure 10. An SEM image of 10μm x 10μm squares spaced 20 μm, each implanted using a scanned, 250 nm diameter, 60 keV Au beam to a constant fluence. The doses ranged from $1x10^{13}$ in the first square to $3x10^{16}$ Au ions/cm^2 in the last (see text for more detail).

to ensure a uniform dose when the entire square is traversed.

When the discrete stepping of the beam implants the area of the square with a uniform dose determined by the dwell time t_d, one can stop at that uniform dose or return the beam to the exact same starting pixel and repeat the bombardment to double the dose. One complete traversal of the square is called a loop, and the time to complete one loop and return to the beginning is a repeat time t_r. By programming t_d, t_r, and the # of loops for a fixed beam current it is possible to vary the dose delivery time over a wide range. One can, of course, vary dose delivery times in other ways.

In the preliminary measurement reported here, scanning electron microscopy (SEM) and atomic force microscopy (AFM) were used to assess the final-stage accumulation of damage in the ion bombarded Ge that leads to the surface transformation by measuring the increase in step heights of the of the ion bombarded areas above the un-bombarded surface. The SEM measurements not only provided qualitative images of how the transformation proceeds with dose (like the top-down images in Figure 9) but edge-on SEM images of cleaved Ge samples also were used to measure the step height increase. However, a more accurate method was to use AFM to measure height increases of each 10μm x 10μm square as a function of dose.

An example of an AFM step height measurement is shown in Figure 11. AFM as well as SEM were also used to measure surface roughness but are not discussed here.

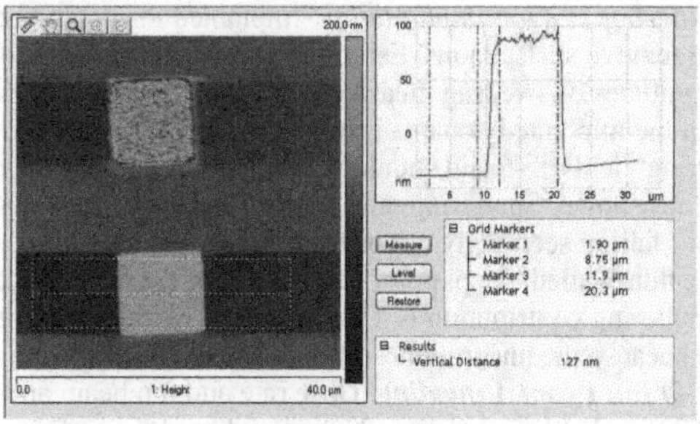

An analysis of the influence of dose delivery times on the measured step heights are summarized in Figure 12 where step heights are plotted versus the fluence F(ions/cm^2) divided by the dwell time t_d (sec.).

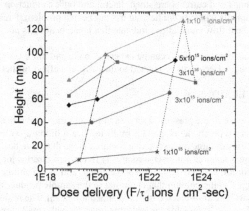

Figure 11. AFM measurement of the increase in step height (swelling) of a 10μm x 10μm area implanted to a uniform dose of 1x10^{16} Au ions/cm^2, relative to un-implanted Ge.

There are three sets of measurements in these preliminary results and they show a definite dependence on the delivery times. The data sets connected by the dashed line and dashed-dot line were bombarded for multiple loops and for t_d = 5x10^{-8} sec and t_d = 5x10^{-2} sec. respectively, to achieve the desired implanted fluences. The remaining data set was taken with one loop per implant but varying t_d from 0.1-4x10^{-3} sec. to achieve the desired fluences. The solid lines in Figure 12 connecting like symbols are from analysis of different 10μm x 10μm squares implanted at identical doses. If there were no extenuating circumstances the measured step heights should be identical within experimental error for like doses, and each of these different dose implants should approximate a straight line of constant step height. Instead they show step heights that increase with increasing dose rate, except for the 3x10^{16} Au ions/cm^2 implants.

Figure 12. Height increases above the unimplanted Ge surface measured for five different implanted fluences taken at different dwell times t_d and number of loops, compared as a function of dose delivery (Fluence/ t_d). See text for details.

As mentioned earlier there are a number of factors that could influence these measurements and/or contribute to measured step height differences.

1) Sputtering and ion implantation. Implanted ions could contribute to swelling, and sputtering should remove surface ions. Estimates show that implantation cannot account for the differences or magnitudes of swelling measured, especially when previous measurements show that up to 70% of the implanted dopants is lost when surface pores form [23, 25, 26]. TRIM estimates a Au sputtering yield of ~6 and should remove atoms equally for the fixed dose lines, but not change the trends shown. Sputtering may account for the highest dose line - where the effect should be largest - falling seemingly out of place below the next highest dose in Figure 12. But, given the severe, thin-walled, deep, porous structures of the Ge surfaces it is difficult to predict what the real sputtering contribution is. Consequently, neither of these corrections has been applied in the figure because the uncertainties of both are sufficiently ambiguous to reduce corrections to a guess. *2) Ion Beam Annealing.* Dose rate and ion beam annealing effects have been reported for Si [32,33] and although the production and annealing of damage in Si and Ge are significantly different, and no surface transformation like that in Ge has ever been observed for Si, beam annealing should be considered. But the effects of ion beam annealing in Si are to reduce defects formed by previous damage so any comparable effects in Ge would likely reduce damage and step heights. *3) Beam Heating.* Increased temperatures during ion bombardment could increase step heights (enhance defect mobilities that help form pores), or decrease them (anneal defects so the amorphous phase forms more slowly, or never); the dividing line is around 300 C and arguments can be made both ways. We do not see evidence of melting of the very thin and flimsy pore walls in even our highest dose rate implants, and if one does simple heat flow calculations temperature rises of a few C at most should occur. And since our ion beam dimensions and currents are small, lateral and bulk conduction should be better than for broad area implants where beam heating has not been a problem. Highest dose rate implants, and simple heat flow calculations indicate that beam heating is not a factor, but it cannot be ruled out.

Dose rate damage effects can be expected in Ge and more controlled experiments and early on-set damage effects will be studied to quantify this effect further under MionLiNE conditions.

CONCLUSIONS.

The purpose of the present paper was to present the capabilities of the MionLiNE system that have been recently developed for multi-ion beam lithography and processing, and illustrate its capabilities in three representative research areas that have been pursued during development. Additional features not discussed are; 1) five micro nozzles for injecting gaseous species at the target for ion assisted deposition and etching; and 2) the ability of the system for real-time imaging. The ion column that is normal to the sample produces spatially accurate images from secondary electrons ejected by the scanned ion beam. The multi-ion species can produce large secondary electron yields for real-time imaging with small currents; and have large critical angles for channeling contrast. This integrated system is capable of high performance ion beam lithography, sputter profiling, maskless ion implantation, ion beam mixing, and spatial and temporal ion beam assisted writing and processing over (100 x 100 mm^2) - all with nanometer precision – and is promising for future ion interaction/processing investigations

ACKNOWLEDGMENTS.
This work was supported by the Office of Naval Research (ONR) under Contract Number 00075094 (BA).

56

References

1. G.W. Arnold, *J. Appl. Phys.*, **46** (1975) 4466.
2. C.W. White, J.D. Budai, S.P. Withrow, J.G. Zhu, S.J. Pennycook, R.A. Zuhr, D.M. Hembree Jr., D.O. Henderswon, R.H. Magruder, M.J. Yacaman, et.al., *Nucl. Instrum. Methods B* **127/128,** 545 (1997).
3. Zhongning Dai, S. Yamamoto, K. Narumi, A. Miuashita, and H. Naramoto, *Nucl. Instrum. Methods B* **149,** 108 (1999)
4. A. Meldrum, S. Honda, C.W. White, R.A. Zuhr, L. A. Boatner, *J. Mater. Res.* **16,** 2670 (2001).
5. A. Meldrum, R. Lopez, R.H. Magruder, L.A. Boatner, and C.W. White, *Topics Appl. Physics* **116,** 255 (2010).
6. A. Meldrum, A. Hryciw, K.S. Buchanan, A.M. Beltaos, M. Glover, C.A. Ryan, J.G.C. Veinot, *Optical Materials* **27,** 812 (2005).
7. Yoshiko Takeda, Oleg A. Plaksin, Haisong Wang, Kenichiro Kono, Naoli Umeda, and Naoki Kishimoto, *Optical Review* **13 No. 4,** 231 (2006).
8. W. Skorupa, R.A. yankov, I.E. Tyschenko, H. Frob, T. Bohme, and K. Leo, *Appl. Phys. Lett.*, **68, No.17,** 2410 (1996).
9. M.E. Castagna, *Physica E* **16** (2003) 547
10. L. Pavesi, Optoelectron. Integration Silicon, *Proc. SPIE* **4997,** 206 (2003).
11. J. Valenta, N. Lalic, J. Linnros, *Opt. Mater.,* **17,** 45 (2001).
12. G. Franzo` , S. Coffa, F. Priolo, C.J. Spinella, *J. Appl. Phys.*, **81,** 2784 (1997).
13. X. Michalet, F. F. Pinaud, L. A. Bentolila, J. M. Tsay, S. Doose, J. J. Li1, G.Sundaresan, A. M. Wu, S. S. Gambhir, and S. Weiss, *Science* **307,** 5709, (2005).
14. Sean Stewart and Guojun Liu, *Chem. Mater.* **11,** 1048 (1999).
15. Weihong Tan, Kemin Wang, Xiaoxiao He, Xiaojun, Julia Zhao,Timothy Drake, Lin Wang, and Rahul P. Bagwe, *Published online in Wiley InterScience* (www.interscience.wiley.com). DOI 10.1002/med. 20003
16. X. Wu, H. Liu, J. Liu, K.N. Haley, J.A. Treadway, J.P. Larson, N. Ge, F. Peale, and M.P. Bruchez, *Nat. Biotech.,* **21,** 41-46 (2003).
17. S. Kim, Y.T. Lim, E.G. Soltesz, A.M. De Grand, J. Lee, A. Nakayama, J.A. Parker, T. Mihaljevic, R.G. Laurence, D.M. Dor, L.H. Cohn, M.G. Bawendi, and J.V. Frangioni, *Nat Biotech* **22**: 93-97 (2004).
18. K.S. Beaty, A. Meldrum, J.F. Franck, K. Sorge, J.R. Thompson, C.W. White, L.A. Boatner, S. Honda, *Mater. Res. Soc. Symp. Proc.* 703, V9.38.1 (2002)
19. A. Meldrum, K.S. Buchanan, A. Hryciw, C.W. White, *Adv. Mater.,* **16,** 31 (2004).
20. Bill R. Appleton, S. Tongay, M. Lemaitre, Brent Gila, Joel Fridmann, Paul Mazarov, Jason E. Sanabia, S. Bauerdick, Lars Bruchhaus, Ryo Mimura, and Ralf Jede, *Nucl. Instr. and Meth.* (in press) doi:10.1016/j.nimb.2011.01.054
21. Brent Gila' Bill R. Appleton' Joel Fridmann',Paul Mazarov, Jason E. Sanabia, S. Bauerdick, Lars Bruchhaus, Ryo Mimura, and Ralf Jede, *CAARI 21: 21ˢᵗ International Conference on the Application of Accelerators in Research and Industry, AIP Conference Proceedings* (2011) in press.
22. A. Lugstein, B. Basnar, E Bertagnolli; *JVST,* **B 20(6),** 2238 (2002).
23. B. R. Appleton, O. W. Holland, J. Narayan, O. E. Schow III, J. S.Williams, K. T. Short, and E. M. Lawson, *Appl. Phys. Lett.,* **41,** 711 (1982).
24. I. H. Wilson, *J. Appl. Phys.,* **53,** 1698 (1982).

25. O.W. Holland, B.R. Appleton and J. Narayan, J. Appl. Phys. 54, 2295, (1983)
26. E.M. Lawson, K.T. Short and 3.S. Williams. B.R. Appleton. O.W.Holland, and 0.E. Schow III., *Nucl. Instr. and Meth.,* **209/210,** 303 (1983).
27. B.R. Appleton, *Mat. Resc. Soc. Symp. Proc.,* **Volume 27,** (1984)
28. B.R. Appleton, O.W. Holland, D.B. Poker, J. Narayan and D. Fathy, *Nucl. Instr. and Meth.,* **B 7/8,** 639 (1985).
29. H. Huber, W. Assmann, S. A. Karamian, A. Mücklich, W. Prusseit, E.Gazis, G. Grötzschel, M. Kokkoris, E. Kossionidis, H. D. Mieskes, and R.Vlastou, *Nucl. Instrum. Methods Phys. Res.* B **122,** 542 (1997) and references therein.
30. B. Stritzker, R. G. Elliman, and J. Zou, *Nucl. Instrum. Methods Phys. Res.* **B 175–177,** 193 (2001) and references therein.
31. L. Romano, G. Impellizzeri, M.V. Tomasello, F. Giannazzo, c. Spinella, and M.G. Grimaldi, *J.of Appl. Phys.,* **107,**084314 (2010) and references therein.
32. R.G. Elliman, S.T. Johnson, A.P. Pogany, and J.S. Williams, *Nucl. Inst. and Methods* **B7/8,** 310 (1985) and references therein.
33. O.W. Holland, J. Narayan, and D. Fathy, *Nucl. Inst. and Methods,* **B7/8,** 243 (1985) and references therein.

Mater. Res. Soc. Symp. Proc. Vol. 1354 © 2011 Materials Research Society
DOI: 10.1557/opl.2011.1213

Ion irradiation effects in silicon nanowires

K. Nordlund[1], S. Hoilijoki[1], and E. Holmström[1]
[1] Helsinki Institute of Physics and Department of Physics, P.O. Box 43, FI-00014 University of Helsinki, Finland

ABSTRACT

Ion irradiation effects in nanowires are of increasing interest due to potential applications of the wires as e.g. current-carrying elements in transistors or as efficient light emitters. Although several experiments have already demonstrated such functionalities, very few theoretical studies on the fundamental mechanisms of ion irradiation have been carried out. To shed light on the basic mechanisms of nanowire irradiation, we have simulated 0.03 - 10 keV Ar ion irradiation of Si nanowires with a $< 111 >$-oriented axis and with all side facets being $< 112 >$. We compare the results with those for Si surfaces and bulk. The results show that the damage production in the nanowire is strongly influenced by surface effects.

INTRODUCTION

Ion implantation is a standard method for introducing dopants into semiconductors. While conventional implantation of course deals with irradiation of bulk materials, recent development of transistors using Si nanowires as the active component [1, 2] raise the question of how ion implantation affects this low-dimensional system. Since a nanowire has a huge surface-area-to-volume ratio, one could expect on one hand that sputtering and transmission is much more pronounced than in bulk Si, on the other hand that the surface might destabilize the system.

Radiation effects in nanowires have been studies to some extent experimentally (for a review see [3]), and major differences (such as a much larger radiation hardness in some nanowire materials) to bulk materials have been reported. In the particular case of Si nanowires, it appears that as in its bulk counterpart, also the Si nanowire can be strongly damaged by the irradiation, and annealing is necessary to obtain functional devices. However, the understanding of damage in Si nanowires is very poor. As a step to obtain such understanding, we have carried out molecular dynamics computer simulations of the primary damage states produced by Ar ion irradiation of Si nanowires. To be able to assess whether finite-size effects contribute significantly to the damage production, we also carried out simulations of bulk Si (by Si self-recoils) and Si thin films of the same thickness as the nanowires.

SIMULATION METHOD

We simulated the irradiation using classical molecular dynamics [4] with the PARCAS code [5]. For modeling the Si–Si interactions, the analytical Stillinger-Weber (SW) three-body potential was used [6]. This potential was chosen because recent comparisons of

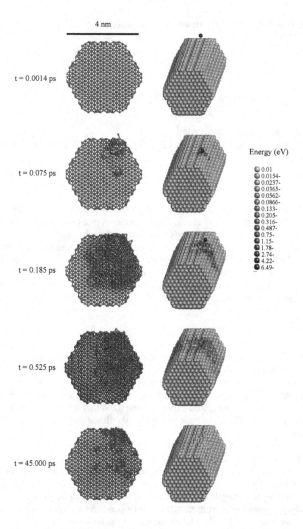

4 nm

t = 0.0014 ps

t = 0.075 ps

Energy (eV)

0.01
0.0154-
0.0237-
0.0365-
0.0562-
0.0866-
0.133-
0.205-
0.316-
0.487-
0.75-
1.15-
1.78-
2.74-
4.22-
6.49-

t = 0.185 ps

t = 0.525 ps

t = 45.000 ps

Figure 1: Snapshots of a 1 keV Ar impact on a Si nanowire. The left side shows a cross section of the whole simulation cell, plotting atoms and covalent bonds. The right side shows exactly the same event at the same time steps in a 3-dimensional view, with a gray scale proportional to the kinetic energy of the atoms.

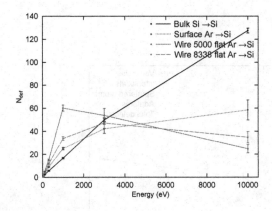

Figure 2: Total defect production in Si bulk, surface(slab) and nanowire irradiations. 5000 stands for the 3 nm and 8338 for the 4 nm wire.

potentials by us showed that, out of the most commonly used empirical potentials for Si, the SW potential gives values for the threshold displacement energies of bulk Si which are closest to experiment and *ab initio* calculations [7, 8]. For the Ar–Si interactions, a purely repulsive ZBL potential[9] was used. Similarly for Si–Si collisions, a repulsive ZBL pair potential was joined to the high-energy part of the Si–Si interactions, as is the standard practice in the field[10, 11]. An electronic stopping power was included in the simulations for the ion as well as all recoil atoms with kinetic energy surpassing 10 eV.

We examined nanowires of effective diameters of about 3 and 4 nm. Both of the wires were ∼ 10 nm Å in length and shared the diamond crystal structure of bulk Si. The axis of the NWs coincided with the ⟨111⟩ crystal direction of the conventional Si unit cell, and the wires had a hexagonal cross section with six [112] side surfaces. The 2 x 7.68 surface reconstruction as predicted by experiment[12] and *ab initio* calculations[13] was taken into account. The motivation for studying this type of NW is that is has been shown to be the most stable small-diameter Si NW [14]. To relax the structures, the wires were annealed slowly from a temperature of 10 to 0 K. The wire of about 4 nm diameter is illustrated in Fig. 1.

To simulate irradiation of a wire of infinite length, periodic boundary conditions were applied along the longitudinal axis of each wire, with 3 Å thick layers fixed at both ends of the wire to prevent the entire system from moving due to momentum transfer from the irradiating ion. To model heat dissipation from the impact region into the rest of the NW, Berendsen temperature scaling[15] to a target temperature of 0 K was applied within 5 Å of the fixed region. In the center the atoms were allowed to evolve freely within Newton's

a)

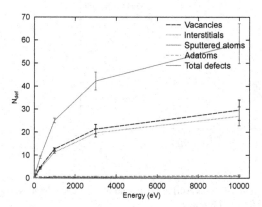

b)

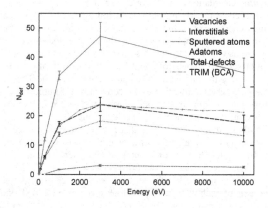

Figure 3: Types of defects produced in a) the surface (slab) simulation case and b) the larger 4 nm wire. Also shown in (b) for comparison is a TRIM simulation of the vacancy production in a Si slab with the same thickness as the nanowire. To make the SRIM results comparable, the threshold displacement energy was fit to reproduce the low-energy MD vacancy production.

equation of motion, except for the friction induced on energetic atoms by the electronic stopping.

Irradiation runs were carried out by irradiating the wires perpendicular to the axis, either against a flat side or an edge. The differences between irradiation against a side or an edge were minor, and will be discussed elsewhere [16]. For each configuration, 200 impact points were chosen randomly within one unit cell along the length direction of the wire. The irradiation simulations were run for 20 ps. The ion energies were 0.03 - 10 keV.

For comparing the results of the NW simulations with the behavior of bulk Si, self-recoils in bulk Si as well as Ar ion irradiation of a reconstructed [001] Si slab surface were simulated using an analogous simulation setup. The simulated slab was chosen to be thick enough so that no damage, sputtering or transmission occurred at the back surface.

Defects were analyzed using Voronoy polyhedron analysis[17]. Moreover, any atom at a distance of 1 to 3 Å from the wire surface was considered an adatom and any atom further than 3 Å from the surface was considered sputtered.

RESULTS AND DISCUSSION

The outcome of a typical 1 keV Ar irradiation event is illustrated Fig. 1, showing that significant amounts of damage are produced both at the wire surface and in the bulk regions of the wire. The total damage production for the simulated cases is quantified in Figure 2. It shows that the total damage production increases with energy in the bulk, as expected from previous works [17]. For the slab case (labeled "surface" in the figure), the damage production is smaller due to sputtering effects. The nanowires, however, show s clear maximum in the damage at 1 keV for the smaller and 3 keV for the larger wire. This is easily understood to be due to ion transmission through the wire, reducing the energy deposition and hence damage production in the wire. A similar effect has been previously reported for carbon nanotubes [18]. A TRIM simulation (see Fig. 3 b) of damage production by Ar in a Si slab of the same thickness showed a similar decrease of damage, confirming that it is indeed a simple issue of ion transmission.

The types of the defects produced in the wires are illustrated in Fig. 3. In the surface case, the vacancy and interstitial production dominate strongly compared to adatom production and sputtering (in the bulk case, there are, of course, no adatoms or sputtering). On the other hand, in the nanowire the fraction of adatoms and sputtered atoms is clearly larger than in the surface case. Moreover, Fig. 2 shows that especially at 1 keV, the damage production in both nanowires is clearly stronger than in bulk Si or Si slabs. Both observations suggest that there may be significant surface effects in the nanowire damage production.

To establish whether surface effects are indeed present, we analyzed the distribution of defects as a function of the defect distance from the center of the nanowire r_{xy}. The results (Fig. 4) show clearly that there is a strong enhancement of the damage production close to the wire surface. This can be understood at least in part based on our earlier observation that the threshold displacement energy in Si nanowires is significantly reduced close to the surface [8, 19].

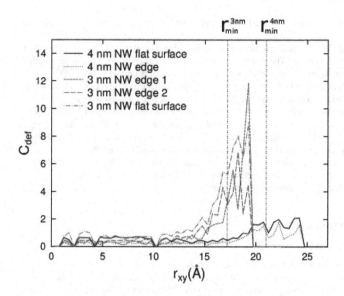

Figure 4: Defect production (interstitials+vacancies) as a function of the defect distance from the center of the nanowire r_{xy} for 1 keV Ar ions. Since the nanowires are not circular in cross section, the distance r_{xy} to the surface varies even within the same effective diameter wire. Hence the figure also shows the minimum distance r_{min} to the surface for both wires.

CONCLUSIONS

We have examined the primary state of damage production in about 3 and 4 nm diameter Si nanowires. At high energies ($\gtrsim$ 5 keV), the damage production in the wires is much smaller than in bulk Si and Si surfaces. This effect can be reproduced by binary collision approximation (SRIM) calculations of thin Si films, showing that it is simply due to ion transmission through the wire. More interestingly, we also find that defect production in the wires is strongly enhanced near the wire surface, showing that the large surface-to-volume ratio in nanoobjects causes special effects also with respect to ion irradiation of nanowires.

ACKNOWLEDGMENTS

This work was performed within the Finnish Centre of Excellence in Computational Molecular Science (CMS), financed by The Academy of Finland and the University of Helsinki. Grants of computer time from the Center for Scientific Computing in Espoo, Finland, are gratefully acknowledged.

REFERENCES

1. S. Hofmann et al., J. Appl. Phys. **94**, 6005 (2003).

2. S. Hoffmann et al., Nano Letters **9**, 1341 (2009).

3. A. V. Krasheninnikov and K. Nordlund, J. Appl. Phys. (Applied Physics Reviews) **107**, 071301 (2010).

4. M. P. Allen and D. J. Tildesley, *Computer Simulation of Liquids* (Oxford University Press, Oxford, England, 1989).

5. K. Nordlund, 2006, PARCAS computer code. The main principles of the molecular dynamics algorithms are presented in [17, 20]. The adaptive time step and electronic stopping algorithms are the same as in [21].

6. F. H. Stillinger and T. A. Weber, Phys. Rev. B **31**, 5262 (1985).

7. E. Holmstrom, A. Kuronen, and K. Nordlund, Phys. Rev. B **78**, 045202 (2008).

8. E. Holmström, A. V. Krasheninnikov, and K. Nordlund, in *Ion Beams and Nano-Engineering, MRS Symposium Proceedings*, edited by D. Ila et al. (MRS, Warrendale, PA, USA, 2009).

9. J. F. Ziegler, J. P. Biersack, and U. Littmark, *The Stopping and Range of Ions in Matter* (Pergamon, New York, 1985).

10. T. Aoki et al., Nucl. Inst. Meth. Phys. Res. B **180**, 312 (2001).

11. T. Aoki, J. Matsuo, and G. Takaoka, Nucl. Inst. Meth. Phys. Res. B **202**, 278 (2003).

12. C. Fulk *et al.*, J. Electr. Mat. **35**, 1449 (2006).

13. C. H. Grein, J. Crys. Growth **180**, 54 (1997).

14. I. Ponomareva, M. Menon, D. Skrivastava, and A. N. Ansdriotis, Phys. Rev. Lett. **95**, 265502 (2005).

15. H. J. C. Berendsen *et al.*, J. Chem. Phys. **81**, 3684 (1984).

16. S. Hoilijoki, E. Holmström, and K. Nordlund, J. Appl. Phys. (2011), submitted for publication.

17. K. Nordlund *et al.*, Phys. Rev. B **57**, 7556 (1998).

18. A. Tolvanen, J. Kotakoski, A. V. Krasheninnikov, and K. Nordlund, Appl. Phys. Lett. **91**, 173109 (2007).

19. E. Holmström, L. Toikka, A. V. Krasheninnikov, and K. Nordlund, Phys. Rev. B. **82**, 045420 (2010).

20. M. Ghaly, K. Nordlund, and R. S. Averback, Phil. Mag. A **79**, 795 (1999).

21. K. Nordlund, Comput. Mater. Sci. **3**, 448 (1995).

Mater. Res. Soc. Symp. Proc. Vol. 1354 © 2011 Materials Research Society
DOI: 10.1557/opl.2011.1278

Folding Graphene with Swift Heavy Ions

Sevilay Akcöltekin[1] , Hanna Bukowska[1] , Ender Akcöltekin[1] , Henning Lebius[2] and Marika Schleberger[1]

[1]University of Duisburg-Essen, 47048 Duisburg, Germany.
[2]CIMAP (CEA-CNRS-ENSICAEN-UCBN), 14070 Caen Cedex 5, France.

ABSTRACT

Swift heavy ion induced modifications on graphene were investigated by means of atomic force microscopy and Raman spectroscopy. For the experiment graphene was exfoliated onto different substrates ($SrTiO_3$ (100), TiO_2(100), Al_2O_3(1102) and 90 nm SiO_2/Si) by the standard technique. After irradiation with heavy ions of 93 MeV kinetic energy and under glancing angles of incidence, characteristic folding structures are observed. The folding patterns on crystalline substrates are generally larger and are created with a higher efficiency than on the amorphous SiO_2. This difference is attributed to the relatively large distance between graphene and SiO_2 of d ≈ 1 nm.

INTRODUCTION

Folding of graphene has attracted enormous interest in recent times. Several groups have studied self-folding or stimulated folding effects of graphene [1-9], either on suspended sheets or on exfoliated flakes. Several possible driving forces for those folding effects have been identified, e.g. ultrasonic excitation [3], chemical interactions [4] or side effects from the preparation. Detailed investigations have shown that the folded parts as well as the edges are mainly oriented along either *zig-zag* or *armchair* directions following the structure of the single layer of carbon atoms arranged in a two-dimensional sp²-bonded honeycomb network [3,10].

Another method to induce folding on single- (SLG) or bilayer graphene (BLG) is the irradiation of graphene by swift heavy ions (SHI) in the kinetic energy range of some tens of MeV [8]. In contrast to the above-mentioned folding techniques SHI irradiation of graphene provides the opportunity to control the folding shape and its dimension and also the number of folded parts. The present paper will demonstrate how to fold graphene by SHI irradiation and discuss on which type of substrate it can be applied.

Generally, if a projectile ion hits a surface one can distinguish roughly between two ion-matter interaction mechanisms: (1) *electronic stopping S_e* which is the dominating mechanism in the higher energy range (> 1 MeV) and described by the initial excitation of the electronic system of the target, (2) *nuclear stopping S_n* denoting the elastic collision of the projectile with the target atoms. Figure 1 shows the stopping power of a typical projectile ion (Xenon) impinging on a $SrTiO_3$ substrate without graphene. The dotted line represents the total stopping power which consists of S_e (solid line) and S_n (dashed line). It can be seen that the second mechanism plays a rather minor role in the SHI energy regime and can therefore be neglected in the following discussion. The typical energy for the projectiles in our experiment was 93 MeV, yielding an electronic stopping power of S_{SrTiO3} ≈ 21 keV/nm in the crystal close to the surface.

Such an energy deposition can finally lead to the production of small single hillocks on the surface of the substrate [12-14].

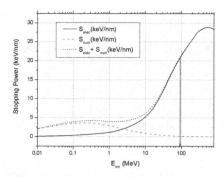

Figure 1 Stopping power as a function of the kinetic energy of a projectile ion impinging on a SrTiO$_3$ (without graphene) crystal, calculated by the SRIM code [15]. Our experiments were performed mostly at 93 MeV (indicated by the black line) which corresponds to a stopping power of about 21 keV/nm.

Especially, irradiation under glancing angles of incidence with respect to the surface may produce long chains of aligned nanosized hillocks with lengths of up to several microns [11]. Based on this idea we apply the same concept to modify graphene prepared on a dielectric substrate.

EXPERIMENTAL DETAILS

For the preparation of the SLG sheets we used the standard technique [16,17] and exfoliated graphene flakes on 10 x 10 mm² polished single crystal insulators such as SrTiO$_3$ (100), TiO$_2$(100), Al$_2$O$_3$(1102) and 90 nm SiO$_2$/Si. In order to make sure that the prepared graphene sheets on the substrates are sufficiently clean and suitable for the following irradiation experiments all surfaces were checked in ambient conditions by using a DI-3100 atomic force microscopy from Veeco. All AFM images were processed with the Nanotec Electronica SL WSxM software [18]. From the raw data (400×400 data points) only a plane was subtracted. After exfoliation, we checked the quality of the graphene sheets by Raman spectroscopy in order to determine regions with single, bi or few layer graphene using a Jobin-Yvon- LabRam Microscope system. The substrates were excited by laser light with a power of less than 5 mW and a wavelength of 514.5 nm. As shown in Figure 2 the spectra show the characteristic 2D mode shapes for SLG. Prior to irradiation we characterized all samples again by means of AFM to ensure that the number of SLG/BLG flakes on the substrate surface is sufficiently high and widely free of defects. This point is essential for the subsequent irradiation of the surfaces since there are no possibilities to steer the beam onto specific areas of the surface. In order to minimize damage during AFM imaging we operated only in the tapping mode. After quality and position check of graphene flakes all samples were irradiated. The irradiation was performed with a beam of Xe^{23+} ions at a total kinetic energy of 93 MeV and at a fluence of 10^9 ions/cm² which corresponds to a nominal yield of 10 ions per square micron. A rotating sample holder in the irradiation provides control of the angle of incidence with an accuracy of 0.5°. All presented irradiation experiments took place at the IRRSUD beam line of the Grand Accélérator National d'Ions Lourds (GANIL) facility in Caen, France.

RESULTS AND DISCUSSION

Figure 2 shows typical AFM images and the corresponding Raman spectra (inset) taken from an area of $1 \times 1 \ \mu m^2$ before irradiation. From the Raman spectra we found a narrow peak at the position of the 2D mode at around 2700 cm^{-1} which is characteristic for single layer graphene. Moreover, we also noticed the absence of the D mode, which confirms the high structural quality of the graphene single layer.

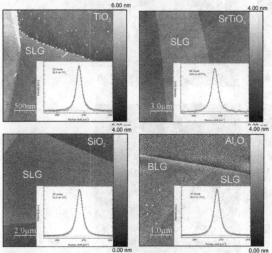

Figure 2 Topography images of graphene exfoliated on different substrates: (a) G/SrTiO₃, (b) G/TiO₂ , (c) G/Al₂O₃ and (d) G/SiO₂. The insets show the corresponding Raman spectra in the region of the 2D peak.

Figure 3 shows the graphene samples after SHI irradiation with Xe ions at a kinetic energy 93 MeV under an angle of incidence of 3°. A comparison between the different substrates covered with graphene reveals similar types of modification. The modifications seem to be stable in air, they could be imaged even several weeks after the irradiation.

The most striking ion induced modifications on each surface in Figure 3 are the triangular shaped folding structures which occur almost exclusively in areas covered with SLG (and with a lower efficiency on BLG, respectively). The orientation of the structures agrees roughly with the direction of the beam which is indicated by the white arrow in Figure 3. From the number of ions per square micron (10 ions) and the number of the observed patterns (9±1 patterns), we estimate that in case of SrTiO₂, TiO₂ and Al₂O₃ each ion impact on SLG has produced one of those structures while on SiO₂ we observed only 4±1 ion-induced defects, corresponding to a reduced efficiency of 40%. The efficiency of folding is further reduced in thicker graphene layers; on FLG we observed no folding effects. However, in FLG we found defect chains oriented along the direction of the incoming ion beam. These structures are very similar to those defects obtained on highly oriented pyrolytic graphite (HOPG) after SHI irradiation under oblique angles [19, 20].

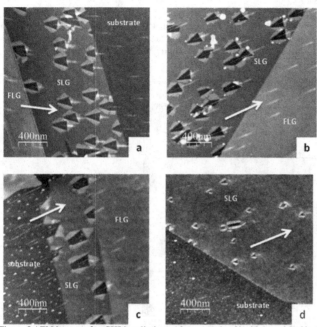

Figure 3 AFM images after SHI irradiation under an angle of incidence of $\Theta=3°$: (a) G/SrTiO$_3$, (b) G/TiO$_2$, (c) G/Al$_2$O$_3$ and (d) G/SiO$_2$. The white arrows indicate the direction of the incoming projectile ions.

A closer inspection of the data reveals that especially the backfolded areas appear higher than the surrounding FLG region. Surprisingly, the folded part is not simply a double layer but appears clearly higher. The reason for this observation is not yet clear. Except for SiO$_2$, the typical folding pattern produced under an angle 3° consists of three backfolded parts with a length between 150 nm and up to 200 nm. In the case of SiO$_2$ defects appear less symmetric and consist mostly of only two folded parts with an average length of 100 nm maximum.

Apart from the folding structure one can observe that each of the folding structures is accompanied by a chain of nanosized hillocks which we call an *ion surface track*. This track is caused in principle by the projectile ion that passed through the surface of the substrate. Due to the high kinetic energy, the projectile ions interact with the surrounding electrons of the substrate along their trajectory. As reported in detail in [21, 22], a consequence of this process is that energy will be transferred via electron-phonon coupling to the lattice so that finally melting and amorphization may occur. From a simple geometric relation which we derived in [23] one can see that the length of such surface tracks is a function of the angle of incidence. At 3°, we find here a length of (193 ± 30) nm for $SrTiO_3$ which is in quite a good agreement with earlier studies (196 ± 29) nm.

These hillock chains can rip the graphene apart, if the tensile strength of graphene is sufficiently weakened by the electronic excitation due to the passing ion as has been discussed in [8]. The striking differences in the folding pattern on SiO_2 compared to the other substrates observed here might thus have a very simple reason. On SiO_2 the average height of SLG is $d \approx 1$ nm [24], while on the other substrates the average height of the graphene is on the order of $d \approx 4$ A [17]. This means in case of SiO_2 the graphene is rather far from the substrate and therefore the pressure exerted by the hillocks is less effective, resulting in a significantly reduced folding efficiency and a different (basically smaller) folding pattern.

CONCLUSIONS

We have shown that SHI irradiation of exfoliated graphene under an angle of incidence of 3° leads to the creation of characteristic folding patterns on dielectric substrates. Depending on the substrate material and the number of graphene layers we have observed only minor deviations in shape, dimension and the number of folding patterns between $SrTiO_3$, TiO_2 and Al_2O_3. However, graphene on SiO_2 exhibits major differences with respect to the shape as well as to the number of observed defects which is roughly half of the expected value. This can be understood if the unusually large distance between graphene and the substrate in case of SiO_2 is taken into account.

ACKNOWLEDGMENTS

We thank the SFB 616: Energy dissipation at surfaces and the EC in the framework of SPIRIT (contract no.: 227012) for financial support. The experiments were performed at the IRRSUD beamline of the GANIL facility in Caen, France.

REFERENCES

1. W. Zhou, Y. Huang, B. Liu, K. C. Hwang, J. M. Zuo, M. J. Buehler and H. Gao, Appl. Phys. Lett. 90, 073107 (2007).
2. N. Patra, B. Wang and P. Kral, Nano Lett. 9 (11), 3766-3771 (2009).
3. J. Zhang, J. Xiao, X. Meng, C. Monroe, Y. Huang and J. Zuo, Phys. Rev. Lett. 104, 166805 (2010).
4. M. J. Allen, M. Wang, S. A. V. Jannuzzi, Y. Yang, K. L. Wang and R. B. Kaner, Chem. Commun., 6285 (2009).
5. S. Cranford, D. Sen and M.J. Buehler, Appl. Phys. Lett. 95, 123121 (2009).
6. Z. Ni, Y. Wang, T. Yu, Y. You and Z. Shen, Phys. Rev. B 77, 235403 (2008).
7. J. H. Warner, M. H Rümmeli, A. Bachmatiuk and B. Büchner, 2010 Nanotechnology 21, 325702 (2010).
8. S. Akcöltekin, H. Bukowska, T. Peters, O. Osmani, I. Monnet, I. Alzaher, B. Ban d'Etat, H. Lebius and M. Schleberger, Appl. Phys. Lett. 98, 103103 (2011).
9. Z. Liu, K. Suenaga, P. J. F. Harris, and S. Iijimal, Phys. Rev. Lett. 102, 015501 (2009).
10. S. Neubeck, Y. M. You, Z. H. Ni, P. Blake, Z. X. Shen, A. K. Geim, and K. S. Novoselov, Appl. Phys. Lett. 97, 053110 (2010).
11. E. Akcöltekin, T. Peters, R. Meyer, A. Duvenbeck, M. Klusmann, I. Monnet, H. Lebius, and M. Schleberger, Nature Nanotechnology 2, 290 (2007).

12. N. Khalfaoui, M. Görlich, C. Müller, M. Schleberger and H. Lebius, Nucl. Instrum. Methods B 245, 246 (2006).
13. A. S. El-Said, M. Cranney, N. Ishikawa, A. Iwase, R. Neumann, K. Schwartz, M. Toulemonde and C. Trautmann , Nucl. Instrum. Methods B 218, 492 (2004).
14. A. Müller , R. Neumann, K. Schwartz and C. Trautmann, Nucl. Instrum. Methods B 146, 393 (1998).
15. J.F. Ziegler and J.P. Biersack, http://www.srim.org, 2008. At the date this paper was written, URLs or links referenced herein were deemed to be useful supplementary material to this paper. Neither the author nor the Materials Research Society warrants or assumes liability for the content or availability of URLs referenced in this paper.
16. A. Geim, K.S. Novoselov, Nature Materials 6 (2007) 183.
17. S. Akcöltekin, M. El Kharrazi, B. Köhler, A. Lorke, and M. Schleberger, Nanotechnology, 20,155601(2009).
18. I. Horcas, R. Fernandez, J. M. Gomez-Rodriguez, J.Colchero, J. Gomez-Herrero, and A. M. Baro, Rev. Sci. Instrum. 78, 013705 (2007).
19. S. Akcöltekin, E. Akcöltekin, H. Lebius and M. Schleberger, J. Vac. Sci. Technol. B 27, 944 (2009).
20. J. Liu, C. Trautmann, C. Müller, and R. Neumann, Nucl. Instr. and Meth. B 193,259 (2002).
21. N. Medvedev, O. Osmani, B. Rethfeld, and M. Schleberger, Nucl. Instrum. Methods B 268, 3160 (2011).
22. O. Osmani, A. Duvenbeck, E. Akcöltekin, R. Meyer, H. Lebius and M. Schleberger, J. Phys.: Condens. Matter 20, 315001 (2008).
23. E. Akcöltekin, S. Akcöltekin, O. Osmani, A. Duvenbeck, H. Lebius and M. Schleberger, New J. Phys. 10, 053007 (2008).
24. E. A. Obraztsova, A. V. Osadchy, E. D. Obraztsova, S. Lefrant and I. V. Yaminsky Phys. Stat. Sol. (b) 245, 2055 (2008).

Mater. Res. Soc. Symp. Proc. Vol. 1354 © 2011 Materials Research Society
DOI: 10.1557/opl.2011.1211

Ion Irradiation on Phase Change Materials

Emanuele Rimini[1], Egidio Carria[2,3], Antonio Massimiliano Mio[2], Maria Miritello[3], Santo Gibilisco[2,3], Corrado Bongiorno[1], Giuseppe D'Arrigo[1], Corrado Spinella[1], Francesco D'Acapito[4], Maria Grazia Grimaldi[2,3].
[1]IMM-CNR, Catania, Italy.
[2]Dipartimento di Fisica e Astronomia, Università di Catania, Italy.
[3]MATIS-IMM-CNR, Catania, Italy.
[4]ESRF GILDA CRG, CNR, IOM OGG, F-38043 Grenoble, France

ABSTRACT

Ion irradiation with 130 keV Ge^+ or 120 keV Sb^+ has modified, by thermal spike effect, the local atomic arrangement in the structure of as-deposited sputtered amorphous GeTe and $Ge_2Sb_2Te_5$ thin films. The changes in the local order have been analyzed by Raman and EXAFS spectroscopy. In addition the crystallization kinetic, measured by "in situ" time resolved reflectivity and optical microscope analysis, is found to be faster in the irradiated samples. The nucleation rate and the grain growth velocity are enhanced by a factor of about ten with respect to the unirradiated samples in the investigated temperature range (120°C -170°C).

INTRODUCTION

Phase-Change Materials (PCMs), mainly represented by $GeTe-Sb_2Te_3$ alloys, are used for high-density data storage in optical media and for solid-state non volatile memory [1-2]. The working principle of these devices is based on the change of the optical and electrical properties during the switch from the amorphous to the crystalline phase and vice versa. Two states are distinguishable by the pronounced difference in reflectivity (up to 30%) and resistivity (some orders of magnitude) between these two phases. Amorphous structure, of relevance for the crystallization kinetics and for the stability of the recorded data, depends on the method adopted for the preparation. Different crystallization behaviour has been found between as deposited, melt quenched and ion irradiated amorphous structures, indicating different local order arrangements [3-4-5]. Among the several procedures ion irradiation offers a unique and well reproducible method to create an amorphous structure. In addition, atomic displacements, bond breaking and bond formation caused by the collision cascade event provide information on the rearrangement in a glassy system as it is a chalcogenide alloy. In this work we report data on the crystallization kinetics of as-deposited and ion irradiated GeTe and $Ge_2Sb_2Te_5$ thin films to be correlated with the local order analyzed by micro-Raman and EXAFS spectroscopy.

EXPERIMENTAL DETAILS

Thin GeTe and $Ge_2Sb_2Te_5$ films, 50 nm thick, were deposited at room temperature by RF-magnetron sputtering over a thermally grown 550 nm thick SiO_2 layer on a Si substrate. Some samples were then covered with a gold circular mask (thickness ~ 25μm and diameter ~ 3mm). Irradiation of as-deposited amorphous films was performed at RT with 130 keV Ge^+ or 120 keV Sb^+ at fluence in the range $1x10^{14} - 1x10^{15}$ ions/cm². The samples were implanted through the mask to make more reliable the comparison between irradiated and unirradiated regions. The beam current was kept constant to 100 nA to avoid heating of the samples. No capping layer was adopted in the present experiment. The effect of sputtering and the influence of an overlayer through recoil implantation were minimized so that no appreciable change of stoichiometry and impurity introduction occur [6]. The stoichiometry and the thickness of amorphous film were evaluated by 2 MeV He^+ Rutherford backscattering spectrometry and particle induced X-ray emission. The crystallization kinetics was investigated by "in situ" time resolved reflectivity using a He-Ne laser probe during annealing from RT to 250°C. The local order and the morphology of amorphous and crystallized samples were detected by micro-Raman, EXAFS, optical and transmission electron microscopy.

RESULTS AND DISCUSSION

A striking illustration of the effect of ion irradiation on the crystallization phenomenon of a GeTe thin film annealed at 153°C is shown by the sequence of optical images reported in Fig.1 [7]. The analyzed region spans an as deposited amorphous area and a contiguous irradiated amorphous area. The enhancement of both nucleation rate and grain growth velocity in the irradiated area is clearly visible as well as the inhibition of grain growth at the border of the irradiated area.

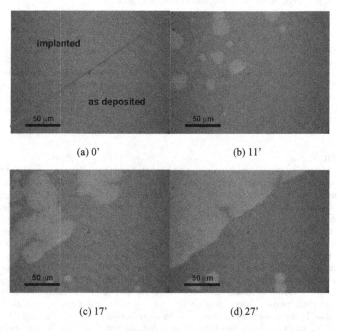

(a) 0'   (b) 11'

(c) 17'   (d) 27'

Figure 1 Optical sequence of as deposited amorphous GeTe sample irradiated trough a mask at a fluence of 1×10^{14} ions/cm^2 and annealed at 153°C. The crystallization dynamics is faster in the implanted region and the grain growth velocity slows down at the border of the unirradiated area (dashed line).

From several analyses at different annealing times and temperatures the dependence on temperature of nucleation rate and grain growth in irradiated and unirradiated areas has been obtained and the corresponding data are reported in Fig. 2 and Fig.3 respectively. Both the graphs show that nucleation rate and grain growth follow, in the investigated temperature range, an Arrhenius behaviour. The nucleation rate is enhanced by about one order of magnitude while the growth velocity is enhanced by a factor of about 3. The differences in the extracted activation energies and preexponential factors are not of a simple physical meaning because of the very narrow range of investigated temperatures and of the experimental errors. In spite of this, the enhancement in nucleation rate and growth velocity is clearly shown.

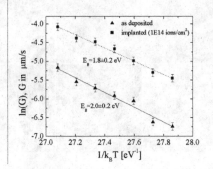

Figure 2 Arrhenius plot of the nucleation rate, I, at several temperatures for the as deposited (▲) and the implanted (■) samples respectively.

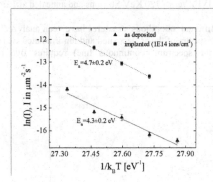

Figure 3 Arrhenius plot of the grain growth velocity, G, at several temperatures for the as deposited (▲) and the implanted (■) samples respectively.

A similar trend has been also observed in $Ge_2Sb_2Te_5$ irradiated and unirradiated amorphous thin films [8]. As an illustration a TEM sequence of micrographs made during "in situ" annealing of irradiated and unirradiated regions is shown in Fig.4. The micrographs of sputtered-deposited films (Fig.4 top) reveal the presence of GST crystal grains whose density and average size increase with the annealing time until their complete coalescence occurs. Transrotational grains, characterized by dark fringes corresponding to bending contour, are dominant in the partially crystallized samples but almost disappear when crystallization is complete. The micrographs in the bottom of Fig.4 represent the morphology evolution in the irradiated regions. Both nucleation rate and growth velocity are enhanced, in addition the grains show very large fringes and a quite uniform contrast, indicating a decrease of the internal bending. The presence of transrotational grains has been attributed to the density difference between amorphous (5.87 g/cm^3) and crystalline (6.27 g/cm^3) GST. As already shown [8] by AFM measurements, ion irradiation of as-deposited amorphous GST and GeTe films induces a densification of the materials and for this reason the amount of transrotational grains is reduced.

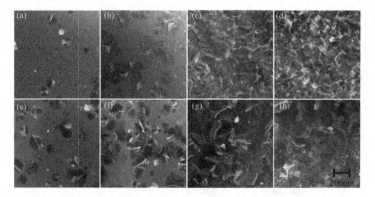

Figure 4. (Top)STEM plan-view micrographs of as-deposited amorphous $Ge_2Sb_2Te_5$ films annealed at 122°C at different times with a crystallized fraction of 20% (a), 40% (b), 80% (c), and 100% (d), respectively.(Bottom)STEM plan-view micrographs of ion irradiated as-deposited amorphous $Ge_2Sb_2Te_5$ films with $1x10^{14}/cm^2$- 120 KeV Sb^+ ions and annealed at 122°C.

The changes in the order of the different amorphous structures have been analyzed by Raman and EXAFS spectrometry. The Raman spectra of the as-deposited and ion irradiated GeTe thin films are shown in Fig. 5. The fit of the spectrum (continuous line) requires five main Gaussian contributions.

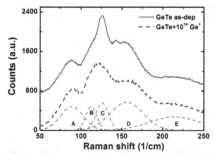

Figure 5. Comparison between the Raman spectra of the unirradiated (solid line) and irradiated (10^{14} ions/cm²) amorphous (dashed line) GeTe thin films. A, B, C, D, and E peaks represent the different vibrational contributions to fit the experimental spectra

The attribution of the peaks to a defined vibrational mode is still under discussion. A recent work by *ab initio* simulations [4] indicates that Ge atoms (mostly 4-coordinated) occupy both tetrahedral and defective octahedral sites while Te atoms (mostly 3-coordinated) occupy only defective octahedral sites. The spectrum above 190 cm^{-1} is associated to tetrahedral structures, while the peaks around 120 and 165 cm^{-1} are mainly due to vibrations of atoms in defective octahedral sites. The peak around 75 cm^{-1} is due to vibrational modes involving threefold coordinates Te atoms. A different description [9], based mainly on a phenomenological model, attributes the peaks to vibrational modes of GeTe$_{4-n}$Ge$_n$ (n=0, 1, 2, 3, 4) building blocks of the GeTe glass structure. It must be pointed out that simulations refer to well defined amorphous structure as obtained by a fast quenching of a liquid phase while the structure of the amorphous material obtained by sputtering is more defective and depends on several parameters. The D band may includes the vibrational mode associated with a disordered arrangement of long Te chains being 157 cm^{-1} the characteristic energy for vibrations of the Te chains in amorphous tellurium. On the other hand short Te chains may contribute to the intensity of the C peak since it is known that crystalline Te has a strong Raman band at 122 cm^{-1}. In addition Te has a very intense Raman cross section. The experimental spectrum of the irradiated amorphous GeTe film shows remarkable changes in the relative intensity of the Raman band C. The atomic mobility induced by the collision cascade causes a reduction of homopolar Te-Te bonds, absent in the crystalline phase, bringing the amorphous system towards a more relaxed state that promotes the crystallization process. Similar trend is also seen in the Raman spectra of as-deposited and irradiated amorphous GST samples.

To gain a better understanding of the changes induced by ion irradiation, some GST films were analyzed by EXAFS. Figure 6 shows the Fourier-transformed spectra, measured at the K-edge of Ge, for as deposited and ion implanted GST. Both Ge-Ge and Ge-Te bonds are required to fit properly the peak due to the first coordination shell and the bond length is 2.51Å and 2.62Å, respectively.

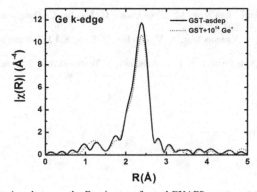

Figure 6. Comparison between the Fourier-transformed EXAFS spectra, at the Ge K-edge, of the unirradiated (solid black line) and irradiated amorphous (dotted red line) GST thin films.

After ion irradiation we observe a damping of the EXAFS oscillations and then a reduction of the peak intensity in the Fourier-transformed signal. This feature is attributed to an increase of the Debye-Waller term after irradiation. A similar behavior has been reported during the crystallization process since the local order of these alloys is higher in the amorphous phase than in the crystalline one [10]. This observation suggests again that after ion irradiation the GST amorphous network is promoted to a state closer to the crystalline phase enhancing the crystallization kinetic.

CONCLUSIONS
In conclusion, we have shown the effect of ion irradiation on the local order and on the crystallization kinetic of amorphous GeTe and Ge$_2$Sb$_2$Te$_5$ thin films. Raman analysis shows the reduction of Te-Te wrong bonds whereas EXAFS data suggest a rearrangement of the Ge-Ge and Ge-Te bonds. The amorphous network is then promoted to a state closer to the crystalline phase. We can then explain the enhanced crystallization kinetic observed in ion implanted samples. After

irradiation, we have also observed a densification of the materials and then a reduced amount of transrotational grains characteristic of GST samples annealed near the glass temperature. Also this phenomenon may enhance the crystallization rate since it is known that densification is involved during the crystallization process.

REFERENCES

[1] M. Wuttig, N. Yamada, Nat. Mater. 6 (2007) 824-832.
[2] S. Lai, IEDM Technical Digest, 2003, pp. 10.1.1-10.1.4.
[3]A.V.Kolobov, P. Fons, J. Tominaga, A. L. Ankudinov, S. N. Yannopolous, K. S. Andrikopolous, J. Phys. Condens. Matter 16 (2004) 55103.
[4] R. Mazzarella, S. Caravati, S. Angioletti-Uberti, M. Bernasconi, M. Parrinello, Phys. Rev. Lett. 104 (2010) 085503.
[5] J. Akola, F03 EPCOS, available on: http://iffwww.iff.kfa-juelich.de/jones/EPCO-Salj_10.pdf, (2010).
[6] E. Carria, A. M. Mio, S. Gibilisco, M. Miritello, M. G. Grimaldi and E. Rimini, Electrochemical and Solid-State Letters, 14 (2011) H124-H127.
[7] A. M. Mio, E. Carria, G. D'Arrigo, S. Gibilisco, M. Miritello, M. G. Grimaldi, E. Rimini, J. Non-Cryst.Solids (2011).
[8] E. Rimini, R. De Bastiani,, E. Carria, M. G. Grimaldi, G. Nicotra, C. Bongiorno, and C. Spinella, J. Appl. Phys. 105, 123502, (2009).
[9] K. S. Andrikopoulos, S. N. Yannopolous, A. V. Lolobov, P. Fons, and J. Tominaga, J. Phys. Chem. Solids 68, 1074 (2007).
[10] A. Kolobov, P. Fons, A. I. Frenkel, A. L. Ankundinov, J. Tominaga, and T. Uruga, Nature Mater. 3, 703 (2004)

Mater. Res. Soc. Symp. Proc. Vol. 1354 © 2011 Materials Research Society
DOI: 10.1557/opl.2011.1216

Ion Beams for synthesis and modification of nanostructures in semiconductors

Anand P. Pathak[*], N. Srinivasa Rao, G. Devaraju, V. Saikiran and S. V. S. Nageswara Rao
School of Physics, University of Hyderabad, Hyderabad 500046, A P, India

ABSTRACT

Swift heavy ion irradiation is one of the most versatile techniques to alter and monitor the properties of materials in general and at nanoscale in particular. The materials modification can be controlled by a suitable choice of ion beam parameters such as ion species, fluence and incident energy. It is also possible to choose these ion beam parameters in such a way that ion beam irradiation can cause annealing of defects or creation of defects at a particular depth. Here, we present a review of our work on swift heavy ion induced modifications of III-V semiconductor heterostructures and multi-quantum wells in addition to synthesis of Ge nanocrystals using atom beam co-sputtering, RF magnetron sputtering followed by RTA, swift heavy ion irradiation, respectively. We also present the growth of GeO$_2$ nanocrystals by microwave annealing. These samples were studied by using XRD, Raman, PL, RBS and TEM. The observed results and their explanation using possible mechanisms are discussed in detail.

----

*Corresponding author. Tel.: +91 40 23010181/23134316; fax: +91 40 23010181/23010227.
E-mail address: appsp@uohyd.ernet.in (A.P. Pathak).

INTRODUCTION

In III-V compounds, introduction of another column III or column V element leads to small change in lattice parameter accompanied by corresponding variation in energy band gap of original semiconductor [1]. For example introduction of small amounts of Indium or Aluminum in GaAs or GaN would result in a change of lattice parameter and band gap. If such layers are deposited on the parent substrate, the small lattice mismatch induces tensile or compressive strain in the deposited epilayer. This strain increases with thickness of the deposited layer or with concentration of the doped elements. Beyond a certain critical value, the strain relaxes into misfit defects. Such misfit dislocations and interfaces play a crucial role in the performance of strained heterostructure devices. Several researchers have investigated the nature of these dislocations and the mechanism of their formation [1–2]. Dislocation formation due to strain relaxation beyond a critical layer thickness is a limitation in strained heterostructures. To overcome this limitation several buffer structures have been proposed [3–4], which proved to be the most efficient solution. The optoelectronic device performance deteriorates drastically due to generation of these misfit defects [5-6]. This leads to possibility of fabricating Strained Layer Superlattice (SLS) with equal thickness of alternate layers of small lattice mismatch with thickness below critical value for strain relaxation. The usefulness of these structures is that, they offer precise control over the states and motions of charge carriers in semiconductors [1]. As a result, the electronic and optoelectronic properties are enhanced many-fold. The strain produced in SLS improves the device performance and is one of the important parameters for tailoring the band structure. Spatial band gap tuning of such Heterostructures (HS) is also important because the integration of photonic circuits demands different band gaps for different devices. Meeting such band gap requirements is quite difficult during the growth. The alternative way then would be to alter the band gap after growing the structures. Compositional disordering and mixing at the interface by ion implantation and subsequent thermal annealing is normally employed. Swift heavy ion (SHI) beam induced mixing is more suitable for the integration of optoelectronic devices because of the advantage that the mixing can be confined to a narrow region at the interface, as compared with the lateral straggling effects in low energy ion irradiation. Monolithic integration of optoelectronic circuits is crucial for fabricating stable photonic devices with submicron alignment of various optical components. These periodic structures with alternating small and large energy band gap lead to realization of Multi Quantum Wells (MQW) offering ideal situations for quantum confinement of conduction/valence electrons at nanoscale. Such structures have in fact been extensively investigated using ion beams, both for their modifications as well as characterizations. Fair amount of work has been done in ion beam treatment of GaAs and GaN based single layers as well as MQW [7-14]. The observed results were interpreted on the basis of existing thermal spike model of high energy heavy ion interaction with target materials. Currently this energy band gap engineering work is being continued and extended to AlGaN and AlInN materials [15, 16] in our group. We have also synthesized and investigated Ge nanocrustals using various established techniques such as such as RF co-sputtering [17], DC sputtering [18] and ion implantation [19]. In addition, we also report on **new** sputtering deposition method namely atom beam sputtering followed by annealing and/or swift heavy ion irradiation to fabricate Si and Ge nano crystals [20-22]. A relatively newer technique, microwave annealing, has also been used in this context for synthesis of Ge and GeO$_2$ quantum dots [23, 24]. By optimizing the parameters this has also been extended to synthesize Si nanocrystals for solar cell applications. The main concern here has been to understand the size and possible modification of the synthesized nanocrystal with fluence and deposited energy by swift heavy ion irradiation. Preliminary results indicate an increase in size of nanocrystal with fluence which seems to be due to agglomeration of smaller nanocrystals. Here we review the work done in our group on these varied materials and techniques with a common goal of synthesizing nanocrystals and characterize these by complementary techniques.

EXPERIMENTAL DETAILS

III-V semiconductor HS and MQW of various combinations (InGaAs/GaAs, InGaAS/InP and GaN) have been grown by using various methods. GaAs based heterostructures were grown by molecular beam epitaxy (MBE) at Solid State Physics Laboratory (SSPL), Delhi. Metal organic chemical vapor deposition (MOCVD) technique was used to grow InP and GaN based MQW at IEMT, Warsaw, Poland. Growth details can be found from the respective references [7-9, 15, 16, 20-25]. Ge and SiO$_2$ composite films were deposited by atom beam sputtering (ABS) and RF magnetron sputtering techniques. As-deposited atom beam sputtered Ge+SiO$_2$ composite films were annealed using rapid thermal annealing (RTA) and RF sputtered films were irradiated with swift heavy ions. GeO$_x$ films were grown on Silicon substrate using Ge target using Ar +O$_2$ reactive RF sputtering. We have used a RXV6 RTP system (AET Thermal Inc) at 700 and 800^{0}C for 120s in N$_2$ atmosphere (2000 SCCM) for RTA. Microwaves used for annealing the RF sputtered samples were produced using a 1.3 KW, 2.45 GHz single mode applicator. Proper amount of silicon carbide was used as microwave susceptor around the sample within the insulating package to preheat the sample and compensate the heat loss from the sample. The temperature of the sample was monitored by using IR pyrometer. Irradiations were carried out using various ions of different energies and fluencies at room temperature using 15 MV pelletron accelerator at Inter -University Accelerator Centre (IUAC), New Delhi. Low-beam current (0.5– 2 pnA) was maintained to avoid heating of the samples. The samples were oriented at an angle of about 5–7° with respect to the beam axis to minimize channeling effects during irradiation. Rutherford back scattering spectrometry (RBS) measurements were carried out using 2 MeV He ions. Si surface barrier detector (energy resolution of 15 keV) placed at a backward angle of 170° was used to detect the scattered particles. High Resolution X-ray Diffraction (HRXRD) experiments were performed using Philip X'pert system with Cu K_α radiation. Profiles of (004) symmetric scans were recorded in ω-scan (Rocking curve) by optimizing the tilt and azimuthal angles. Glancing Incidence X-ray diffraction (GIXRD) measurements were carried out with CuKα X-rays with λ = 0.154 nm in a glancing angle incidence geometry. Raman scattering measurements were carried out at room temperature in backscattering geometry using a 514.5 nm argon-ion laser beam of 75mW power. The scattered photons were collected using a camera lens (Nikkon) and focusing lens. A double monochromator (model Spex 14018) was used for dispersion and photons were detected using a cooled photomultiplier tube (model ITT-FW 130) operated in counting mode. Photoluminescence studies were carried out at room temperature (295 K) as well as at low temperature (18 K). Samples were excited using a YAG laser (532 nm) and the emitted light was dispersed by 2/3 m McPherson monochromator and detected using LN_2 cooled InAs detector. Atomic force microscopy (AFM) was carried out using SPA400, Seiko Instruments Inc. Transmission Electron Microscopy (TEM) measurements were performed using instrument model FEI TECNAI, G2 F-TWIN D2083 with an electron beam accelerating voltage of 200kV.

RESULTS AND DISCUSSION

Irradiation effects on InGaAs/GaAs samples

The effects of 150 MeV Ag ion irradiation on MBE grown $In_{0.1}Ga_{0.9}As$/GaAs samples have been studied using HRXRD. Strain measurements were performed using HRXRD by determining the angular shift in the layer peak position with respect to that of the substrate peak.

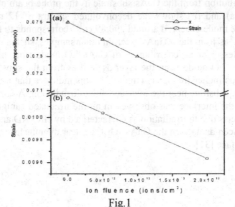

Fig.1

Fig. 1. (a) A plot of In composition as a function of ion beam fluence, (b) a plot of compressive strain as a function of ion beam fluence.

As shown in Fig. 1(a) the concentration of *In* in InGaAs layer decreases with increase in ion fluence due SHI induced diffusion of *In* into the substrate region. Fig. 1(b) shows the plot of strain (compressive) versus the ion fluence, which confirms the compressive strain will decrease with the increase in fluence. It was also shown that SHI can induce tensile strain in initially lattie matched system (InGaAs/InP). Hence high energy ion irradiation can be used to tune the strain without destroying the quality of the material [25].

The $In_{0.18}Ga_{0.82}As$ layers with thicknesses of 12, 36, 60 and 96 nm were grown on GaAs at 500 °C with a growth rate of 0.2 nms^{-1} by MBE. SHI irradiation was performed at room temperature using 150 MeV Ag^{12+} ions at a fluence of 1×10^{13} ions cm^{-2}. The effect of SHI on InGaAs/GaAs HS was studied using Raman spectroscopy. The Raman spectra of the samples are given in figures 2 and 3. InAs type modes were hardly observed, probably due to a lower indium concentration [26, 27]. In the present experimental configuration, the GaAs type TO mode is forbidden. However, a weak TO mode at about 270 cm^{-1} appears due to strain relaxation-induced defects in the samples. The usual intense peak of the GaAs type LO mode around 290 cm^{-1} is also observed [28, 29]. The results are discussed in the light of penetration depth of the probe laser beam used in the experiment. The penetration depth of a 514.5 nm laser beam in GaAs is about 55 nm and in InAs it is about 15 nm in defect free crystals. These values are significantly reduced by disorder present in the crystal [28]. Hence, in the present study, the interface of 12 and 36 nm thick samples is within the penetration depth. However, for 60 and 96 nm thick samples, the interface is beyond the penetration depth of the probe beam. For the unirradiated samples, the strain decreases as a function of thickness; this indicates that the onset of strain relaxation is around 12 nm. Very low strain values for thick samples indicate a strong relaxation of strain in the near-surface regions. After irradiation, a blue shift of the Raman mode, implying an increase of strain in thin layers, and a red shift, implying a decrease of strain in thick layers, were observed. This is probably due to the variation in the ion beam modifications near the interface and near the surface. The 12 nm thick samples have a broad GaAs type LO mode, which is due to the contribution from the GaAs substrate as the probe beam also penetrates into the substrate. Figures 2(a) and (b) shows the deconvoluted spectra of 12 nm thick U and I samples, respectively. The mode observed around 290.89 cm^{-1} corresponds to the bulk GaAs and the mode around 295 cm^{-1} is from the InGaAs layer. In other samples, such a difference in strain was negligible due to a lesser or no contribution of GaAs LO mode from the substrate. After irradiation, the GaAs type LO mode from the layer as well as from the substrate of the 12 nm thick sample shifts towards higher frequency (figure 2). This indicates that a compressive strain has been developed in the substrate near the interface due to irradiation. A compressive strain in the substrate region near the interface was observed in all the irradiated samples using HRXRD [30].The reduction of defects due to irradiation as characterized by HRXRD and RBS/channeling was attributed to ion-induced damage in the GaAs substrate near the interface and inter-diffusion of indium across the interface [31].

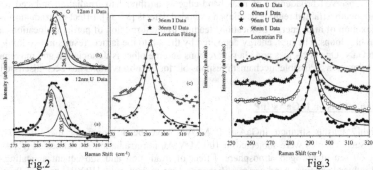

Fig.2

Fig.3

Fig.2 (a) and (b) Deconvoluted Raman spectra of 12 nm thick U and I samples, respectively, and (c) Raman spectra of 36 nm U and I samples. Fig.3 Raman spectra of 60 and 96 nm thick U and I samples.

Irradiation effects on GaN

Epitaxial GaN layers grown by MOCVD on c-plane sapphire substrates were irradiated with 150 MeV Ag ions at a fluence of 5 x 10^{12} ions/cm². Samples used in this study are 2 μm thick GaN layers, with and without a thin AlN cap-layer. The residual strain and sample quality have been analyzed before and after irradiation using HRXRD. Optical properties were studied by spectrophotometer in transmission mode. The spectral transmittance curves (Fig. 4) show sharp absorption edge for as grown samples. In contrast, irradiated sample shows increase in band-edge absorption with photon energy. This kind of behavior is observed in highly defected samples or at high doping level. Band edges are calculated using $(\alpha E)^2$ versus hv plots shown in Fig. 5. In the as grown samples measured band gaps are 3.39, 3.39 and 3.38 eV for RE48U, A4U and A6U, respectively. After irradiation the calculated band gap values are decreased to 3.34, 3.24 and 3.29 eV for RE48I, A4I and A6I, respectively.

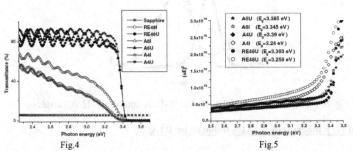

Fig.4

Fig.5

Fig. 4. Spectral transmission curves of the as grown and irradiated samples. Fig. 5. Band gap values derived from optical transmission data for as grown and irradiated samples.

The observed broadening of near band-edge is attributed to irradiation induced point defects and locally strained regions or relaxed extended defects [32-34]. Effect of irradiation is to induce relaxation of high energy dislocations leading to the formation of partial dislocations and dislocation half loops. The high energy deposited by the ion to the lattice, results in two kinds of damage in GaN. One is N_2 loss and Ga rich regions and the other is dissociation and movement of stable dislocations. Possible mechanism responsible for the observations has been discussed in Ref [35].

Irradiation effects on InGaAs/InP

Highly tensile strained InGaAs/InP MQWs have been grown by the LP-MOVPE technique. Such samples were subjected to 100 MeV Au ion irradiation and annealed (RTA) at 700°C for 60 sec in nitrogen atmosphere. Effect of irradiation and subsequent annealing on interfacial induced changes, band gap modifications were studied using HRXRD and PL. Pristine MQWs, irradiated with 5 x 10^{12}, 1 x 10^{13} ions/cm^2 and followed by RTA at 700 °C are given sample IDs as MQW-U, MQW-I1-A and MQW-I2-A, respectively. Fig. 6 shows LT-PL spectra of MQW-U, MQW-I1-A and MQW-I2-A samples. The irradiated MQWs do not show any PL due to irradiation induced non-radiative recombination centers. LT-PL was observed on irradiated and subsequently annealed samples. This study reveals that the band gap can be tailored by irradiation with different ion fluence due to increase in tensile strain induced by the interfacial atomic mixing. The sample MQW-I1-A has good PL intensity and a better peak width compared to MQW-U. The peak profile of MQW-I2-A has two Gaussian profiles indicating residual defects. We showed that band gap can be tailored to 47 nm and 82 nm for MQW-I1-A, MQW-I2-A samples [36], respectively.

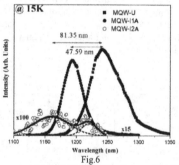

Fig.6

Fig. 6. LT-PL spectra of MQW-U, MQW-I1-A and MQW-I2-A samples

Synthesis of Ge nanocrystals by ABS followed by RTA

Thin films of Ge and SiO_2 nanocomposites were deposited by ABS of Ge and silica with fast neutral Ar atoms having an energy of ~1.5 keV. These as deposited films were annealed at 700 and 800°C for 120s in N_2 atmosphere (2000 SCCM). Measured and simulated RBS spectra of these Ge and SiO_2 composite films are shown in Fig.7. The film thickness and the concentration of Ge in this film were estimated to be about 140 nm and 32 at% respectively. Fig. 8 shows the RBS spectrum of annealed sample at 800°C. The Ge concentration in this film is around 28 at%. The observed loss of Ge in 800°C annealed sample is attributed to partial out-diffusion or evaporation of Ge from the surface. GIXRD spectra of set A as-deposited and RTA samples are shown in Fig.9. In the case of the as-deposited sample, no peak is observed, whereas the annealed samples at temperatures 700°C and 800°C show three peaks of Ge (111), (220) and (311) planes indicating the formation of nc-Ge [37].

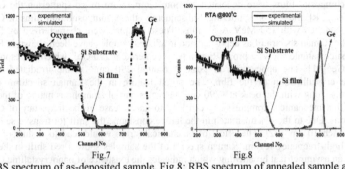

Fig.7   Fig.8

Fig.7: RBS spectrum of as-deposited sample, Fig.8: RBS spectrum of annealed sample at 800°C

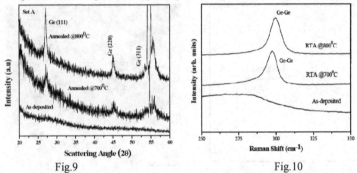

Fig.9   Fig.10

Fig. 9: GIXRD spectra of as-deposited and annealed samples, Fig.10: Raman spectra of as-deposited and annealed samples

The samples annealed at 700°C and 800°C indicate that the average diameter of the nanocrystal is around 12 and 21 nm, respectively which is estimated by using Scherrer formula. Thus, the size of the nc-Ge increases with annealing temperature. This is in good agreement with the results reported in the literature [38]. Raman spectra of rapid thermal annealed samples are shown in Fig 10. Raman spectra of all the samples show a shift in the peak position and an asymmetrical broadening on the lower frequency side when compared to that of bulk Ge sample. The as-deposited sample shows a broad Raman peak centered at 270 cm^{-1} which corresponds to amorphous Ge [39]. The appearance of sharp peak around 297 cm^{-1} for sample annealed at

700°C indicates the initiation of crystallization process. Raman spectra of the samples show red shift in the peak position and an asymmetrical broadening when compared with the spectrum of the bulk Ge sample. This broadening of the Raman peak is attributed to phonon confinement [40]. Growth of the size of Ge nanocrystals with annealing temperature and advantages of ABS over other methods were discussed in ref [20].

Synthesis of Ge nanocrystals by RF Sputtering followed by SHI irradiation

The composite films were deposited at room temperature by co-sputtering the Ge and SiO_2 target using RF magnetron sputtering. Subsequently, these composite thin films of Ge and silica were irradiated with 150 MeV Ag^{+12} ions. These samples were characterized by Raman and TEM. The Raman spectra of as deposited and irradiated samples have been shown in Fig.11. The as-deposited sample shows a broad Raman peak centered at 270 cm^{-1} which corresponds to amorphous Ge [41]. The appearance of peak at around 290 cm^{-1} in the irradiated sample indicates the crystallization of the film. The broad feature in the Raman spectrum of the irradiated films along with the peak at ~290 cm^{-1} indicates partial amorphous nature of the film. It is known that mechanical compressions can lead to an increase in the frequency of Ge–Ge bond vibrations. Due to the 4% mismatch in the lattice constants, the shift for transverse optical modes can reach 10 cm^{-1}[42]. The effect of mechanical stress would lead to a shift of peaks towards the high-frequency region. Raman spectra of the samples show a red shift in the peak position and an asymmetrical broadening which indicates the existence of nanocrystalline Ge.

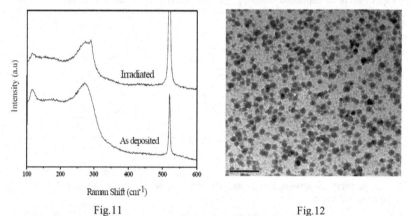

Fig.11   Fig.12

Fig.11: Raman spectra of as deposited and irradiated samples, Fig.12: TEM image of 150 MeV Ag irradiated sample with fluence of 3 x 10^{13} ions/cm^2

The formation of nanocrystallites as a result of Ag ion irradiation was also observed from TEM micrographs as shown in Fig.12. The average nanocrystallite size is around 20 nm. It is possible that Ge nanocrystals with different average sizes can be synthesized by varying ion fluence and energy. One can explain the basic mechanism of recrystallization and formation of nanocrystals in these samples, under ion-irradiation with the help of thermal spike theory [43]. When the swift heavy ion passes through the material the energy is transferred to the target lattice via electron–phonon coupling. It is known from the work of Meftah et al. [44] and Toulemonde et al. [45] that a track diameter of about 10 nm is created for a S_e value of ~11 keV/nm in silica. This large amount of energy is transferred to the target electrons in very short time, in the order of ~10^{-12} seconds which leads to an increase of the lattice temperature above its melting point along the ion track. The temperature along the ion path goes to very high value which is sufficient to crystallize the Ge that is embedded in Silica. This results in a modification of structure of films around the cylindrical zone and crystallization of the films [46].

Synthesis of GeO₂ nanocrystals

GeO$_x$ thin films have been deposited on silicon substrate using RF reactive magnetron sputtering with Ge target using Ar and O$_2$ as sputtering and reactive gases respectively. The as-deposited samples were annealed at 900°C using microwave annealing. All the samples were subsequently characterized by XRD to observe the formation of GeO$_2$ nanocrystals. Raman spectroscopy measurements were also carried out to confirm the presence of the nanocrystals. Fig. 13 shows the XRD spectrum of the GeO$_x$ as deposited and annealed film at 900°C using microwave annealing. All the diffraction peaks including (100), (011), (110), (102), (111), (200), (021), (003), (112), (013), (202), (210) and (211) can be clearly indexed to the hexagonal structure of GeO$_2$. The calculated lattice constants of GeO$_2$ from experimental data, a = 4.982 and c = 5.644 Å are in close agreement with JCPDS tables where a = 4.9858 Å and c = 5.6473 Å. The size of the nanocrystal was estimated to be around 40 nm using Scherrer formula. All the peak positions are in good agreement with the expected JCPDS table values.

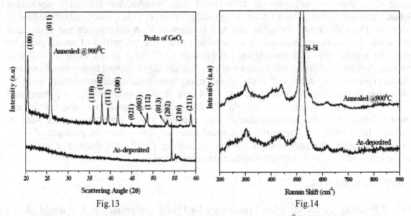

Fig.13   Fig.14

Fig.13: GIXRD spectra of as-deposited and annealed samples at 900°C, Fig.14: Raman spectra of as-deposited and annealed samples at 900°C

Raman spectra of GeO$_x$ pristine and annealed films have been shown in Fig.14. The peak at 521 cm^{-1} is attributed to the optical phonon mode of the Si substrate [47].The peak at 442 cm^{-1} in the as deposited sample shows a broad feature which may be due to amorphous nature. However, the sharp peak at 443 cm^{-1} in annealed films is related to the crystalline GeO$_2$ [48].

Besides, we observe a peak at 301 cm^{-1}, which is due to Ge optical phonons. Most of the time, preparation of GeO$_x$ films using reactive sputtering method yields amorphous films with a composition close to GeO$_2$. Formation of GeO$_2$ nanocrystals in the present system is due to microwave annealing which was carried in air at 900^{0}C. XRD spectra also show the amorphous nature of as deposited films but it clearly indicates the formation of nanocrystals as a result of microwave annealing. Eventually, it is understood that GeO$_x$ to GeO$_2$ transformation takes place due to annealing as discussed above. It is possible to tune the properties of nanocrystal by changing deposition parameters like RF power, Ar, O$_2$ flow rates and distance between target to substrate. One can also change the microwave annealing parameters to alter the structural and optical properties of the nanocrystals. The results obtained here vis-à-vis the importance of microwave annealing have been discussed in detail [24].

CONCLUSIONS

In this review paper, we have collated our work on ion beam synthesis, modification and characterization of nanosized semiconductors. It is found that the compressive strain normally decreases with the increase in ion fluence. This further implies that high energy ion irradiation may be used to tune the strain without destroying the quality of the material in InGaAs/GaAs samples. Strain modification of SHI-irradiated InGaAs/GaAs HS using Raman spectroscopy has been reported for the first time. The Raman results are discussed in the light of the penetration depth of the probe beam. The high energy deposited by the ion to the lattice, results in two kinds of damage created in GaN. One is N loss and Ga rich regions, and the otheris dissociation and movement of the stable dislocations. It is observed in InGaAs/InP MQWs that non-radiative recombination vanishes for irradiated samples after annealing. The irradiation and subsequent annealing parameters are necessary to tailor band gap. Besides, atom beam co-sputtering deposition technique followed by RTA has been used to synthesize Ge nanocrystals embedded in SiO$_2$ matrix. The variation of crystallite size as a function of RTA temperature and advantages of ABS method have been discussed. It is possible to tune the properties of nanocrystals by changing the various pre- and post deposition parameters in order to use them for different applications. Synthesis of Ge nanocrystals by using RF sputtering followed by ion irradiation has also been reported. As deposited films show amorphous nature but irradiation leads to crystallization of the films as evidenced by TEM and Raman spectroscopy. We have also synthesized GeO$_2$ nanocrystals on silicon by RF magnetron sputtering and subsequent microwave annealing. XRD indicates that as-deposited film shows amorphous nature whereas the annealed films show crystalline nature. Raman spectrum also shows the presence of GeO$_2$ nanocrystal formation after performing microwave annealing. Finally it is shown that ion beams can be used to synthesize and alter semiconductor nanostructures.

ACKNOWLEDGMENTS

A P P thanks Center for Nano Technology for DST-Nano project. N S R would like to thank UGC DAE-CSR for fellowship. G D and V S would like to thank CSIR for SRF. SVSNR thanks MRS for travel support to attend the MRS meeting. We would like to acknowledge, Dr R Muralidharan, SSPL, Delhi, Dr D K Avasthi, IUAC, New Delhi, Prof B M Arora and Prof Arnab Bhattacharya, TIFR, Mumbai, Dr K G M Nair and Dr B Sundaravel, IGCAR, Kalpakkam, C Jagadish, ANU Canberra, Dr M Ghanashyam Krishna and Dr K C James Raju, UoH for their collaboration during various aspects of the work reviewed here.

REFERENCES

1. Semiconductors and Semimetals, Vo. 33, Ed. T. P. Pearsal, Acadmic Press, New York (1991) and ref there in.
2. A.V. Drigo, A. Aydinli, A. Carnera, F. Genova, C. Rigo, C. Ferrari, P. Fanzosi, G. Saviati, *J. Appl. Phys.* **86**, 1975 (1989).
3. J. Tersoff, *Appl. Phys. Lett.* **65**, 2460 (1993).
4. D. Araujo, D. Gonzalez, R. Garcia, A. Sacedon, E. Calleja, *Appl.Phys. Lett.* **67**, 3632 (1995).
5. J.W. Mathews, A.E. Blakeslee, *J. Cryst. Growth* **27**, 118 (1974).
6. I.J. Fritz, S.T. Picraux, L.R. Dawson, T.J. Drummond, W.D. Lsidig, N.G. Anderson, *Appl. Phys. Lett.* **46**, 987(1985).
7. S. Dhamodaran, N. Sathish, A.P. Pathak, D.K. Avasthi, R. Muralidharan, B. Sundaravel, K.G.M. Nair, D.V. Sridhara Rao, K. Muraleedharan and D. Emfietzoglou, *Nucl. Inst. Meth. B*, **254**, 283 (2007).
8. S. Dhamodaran, A.P. Pathak, A. Turos, R. Kesavamoorthy, B. Sundaravel, K.G.M. Nair and B.M. Arora, *Nucl. Inst. Meth. B*, **266**, 1908 (2008).
9. S. Dhamodaran, A.P. Pathak, A. Turos, G Sai Saravanan, S.A. Khan, D.K. Avasthi and B.M. Arora, *Nucl. Inst. Meth. B*, **266**, 583 (2008).
10. S.O. Kucheyev, J.S. Williams, S. Pearton, *J. Mater. Sci. Eng. R.* **33**, 51 (2001).
11. S.O. Kucheyev, H.J. Timmers Zou, J.S. Williams, C. Jagadish, G. Li, *J. Appl. Phys.* **95**, 5360 (2004).
12. W. Jiang, W.J. Weber, M. Wang, K. Sun, *Nucl. Inst. Meth. B* **218**, 427 (2004).
13. V. Suresh Kumar, J. Kumar, D. Kanjilal, K. Asokan, T. Mohanty, A. Tripathi, Francisca Rossi, A. Zappettini, L. Lazzarani, *Nucl. Inst. Meth. B* **266**, 1799 (2008).
14. Bolse W and Schattat B, *Nucl. Inst. Meth. B* **190**, 173 (2002).
15. N Sathish, A P Pathak, S Dhamodaran ,B Sundaravel, K G M Nair, S A Khan, D K Avasthi, M. Bazzan E. Trave, P. Mazzoldi, *Radiation Effects and Defects in Solids*. (In Press)
16. G. Devaraju, N. Sathish, A.P. Pathak, A. Turos, M. Bazzan, E. Trave, P. Mazzoldi and B.M. Arora, *Nucl. Inst. Meth. B* **268**, 3001 (2010).
17. P. Caldelas, A.G. Rolo, A. Chahboun A, S. Foss, S. Levichev, T.G. Finstad, M.J.M. Gomes, O. Conde, *J. Nanosci. Nanotechnol.* **8**, 572 (2008).
18. M. Zacharias, P.M. Fauchet, *J. Non-Cryst. Solids*, **227–230**, 1058 (1998).
19. S.N.M. Mestanza, E. Rodriguez, N.C.Frateschi, *Nanotechnology* **17**, 4548 (2006).
20. N. Srinivasa Rao, A.P. Pathak, N. Sathish, G. Devaraju, V. Saikiran, P.K. Kulriya, D.C. Agarwal, G. Sai Saravanan and D.K. Avasthi, *Solid State Communications* **150**, 2122 (2010).
21. N. Srinivasa Rao, S. Dhamodaran, A.P. Pathak, P.K. Kulriya, Y.K. Mishra, F. Singh, D. Kabiraj, J.C. Pivin and D.K. Avasthi, *Nucl. Inst. Meth B* **264**, 249 (2007).
22. N. Srinivasa Rao, A.P. Pathak, N. Sathish, G. Devaraju, S.A. Khan, K. Saravanan, B.K. Panigrahi, K.G.M. Nair and D.K. Avasthi, *Nucl. Inst. Meth. B* **268**, 1741(2010).
23. N. Srinivasa Rao, A.P. Pathak, G. Devaraju and V. Saikiran, *Vacuum*, **85**, 927 (2011).

24. V Saikiran, N Srinivasa Rao G Devaraju and A P Pathak, *AIP conference proceedings*, **1336**, 264 (2011).
25. S.V.S.Nageswara Rao, A. K. Rajam, Azher M. Siddiqui, D. K. Avasthi, T. Srinivasan, Umesh Tiwari, S. K. Mehta, R. Muralidharan, R. K. Jain and Anand P. Pathak. *Nucl. Inst. Meth. B* **212**, 473 (2003).
26. Islam M R, Verma P, Yamada M, Kodama S, Hanaue Y and Kinoshita K *Mater. Sci. Eng. B* **91/92**, 66 (2002).
27. Olego D J, Shahzad K, Petruzzello J and Cammack D *Phys. Rev. B* **36**, 7674 (1987).
28. Brafman O, Fekete D and Sarfaty R *Appl. Phys. Lett.* **58**, 400 (1991).
29. Burns G, Wie C R, Dacol F H, Pettit G D and Woodall J M *Appl. Phys. Lett.* **51**, 1919 (1987).
30. S. Dhamodaran, N. Sathish, A.P. Pathak, S.A. Khan, D.K. Avasthi, T. Srinivasan, R. Muralidharan and B.M. Arora, *Nucl. Inst. Meth. B*, **256**, 260 (2007).
31. S Dhamodaran , N Sathish , A P Pathak , S A Khan , D K Avasthi , T Srinivasan , R Muralidharan , R Kesavamoorthy and D Emfietzoglou, *J. Phys. Condens. Matter* **18**, 4135 (2006).
32. M.O. Manasreh, *Phys. Rev. B* **53**, 16425 (1996).
33. L. Balagurov and P.J. Chong, *Appl. Phys. Lett.* **68**, 43 (1996).
34. W. Rieger, R. Dimitrov, D. Brunner, E. Rohrer, O. Ambacher and M. Stutzmann, *Phys. Rev. B* **54**, 17596 (1996).
35. N. Sathish, S. Dhamodaran, A.P. Pathak, M. Ghanashyam Krishna, S.A. Khan, D.K. Avasthi, A. Pandey, R. Muralidharan, G. Li and C. Jagadish *Nucl. Inst. Meth. B* **256**, 281 (2007).
36. G. Devaraju, S. Dhamodaran, A.P. Pathak, G. Sai Saravanan, J. Gaca, M. Wojcik, A. Turos, S.A. Khan, D.K. Avasthi and B.M. Arora *Nucl. Inst. Meth. B* **266**, 3552 (2008).
37. X.M. Wu, M.J. Lu and W.G. Yao *Surf. Coat. Technol.* **161**, 92 (2002).
38. M.J. Lu, X.M. Wu and W.G. Yao *Mater. Sci. Eng. B* **100**, 152 (2003).
39. Kanakaraju S, Sood A K and Mohan S *Curr. Sci.* **80**, 1550 (2001).
40. P.M. Fauchet and I.H. Campbell, *Crit. Rev. Solid State Mater. Sci.* **14** S79 (1988).
41. Y. Sasaki and C. Horie, *Phys. Rev. B* **47**, 3811(1993)
42. D. A. Orekhov, V. A. Volodin, M. D. Efremov, A. I. Nikiforov, V. V. Ul'yanov, and O. P. Pchelyakov, *Journal of Experimental and Theoretical Physics Letters*, **81**,331 (2005).
43. M. Toulemonde, C. Dufour and E. Paumier, *Phys. Rev. B* **46**, 14362 (1992).
44. Meftah, F. Brisard, M. Costantini, E. Dooryhee, M. Hage-Ali, M. Hervieu, J. P. Stoquert, F. Studer and M. Toulemonde *Phys. Rev. B* **49**, 12457 (1994).
45. M Toulemonde, J. M. Costantini, Ch. Dufour, A. Meftah, E. Paumier and F. Studer, *Nucl. Inst. Meth. B* **116**, 37(1996).
46. N Srinivasa Rao, A.P.Pathak, N.Sathish, G.Devaraju and V.Saikiran *AIP Conf. Proc. 1336, 341 (2011).*
47. R.A. Asmar, J.P. Atnas, M. Ajaka, Y. Zaatar, G. Ferblantier, J.L. Sauvajol, J. Jabbour, S. Juillaget and A. Foucaran, *J. Cryst. Growth* **279**, 394 (2005).
48. G. Kartopu, S.C. Bayliss, V.A. Karavanskii, R.J. Curry, R. Turan and A.V. Sapelkin, *J. Lumin.* **101**, 275 (2003).

Mater. Res. Soc. Symp. Proc. Vol. 1354 © 2011 Materials Research Society
DOI: 10.1557/opl.2011.1212

Raman scattering study of Si nanoclusters formed in Si through a double Au implantation

Gayatri Sahu and D.P. Mahapatra
Institute of Physics, Bhubaneswar, India
Email:dpm@iopb.res.in

ABSTRACT

A sequential two step 32 keV Au implantation and 1.5 MeV Au irradiation technique has been used to synthesize Si nanoclusters (NCs) in Si. The low energy amorphising implantation has been carried out over a fluence range of $(1 - 100) \times 10^{15}$ cm^{-2} while the high energy recrystallizing irradiation fluence was fixed at 1×10^{15} cm^{-2}. Samples were further annealed in air at temperatures between 500° to 950° C for a fixed annealing time of 1 hr and were characterized using Raman scattering at an excitation wavelength of 514.5 nm. Results on as-implanted and irradiated samples indicate formation of strained NCs in the top amorphised layer. Annealing around 500°C has been found to result in strain relief after which the data could be well explained using a phonon confinement model with an extremely narrow size distribution.

INTRODUCTION

Silicon (Si) has often been considered an impractical material for photonic applications because it is an inefficient light emitter. This can be circumvented in going to nanometer length scales [1]. At sizes below the excitonic Bohr radius ($a_B = 4.3$ nm) Si NCs show quantum confinement effect. At finite size, the carrier (electron and hole) wave-functions spread out in k-space, breaking usual crystal momentum selection rule, leading to quasi-direct transitions [2]. This leads to the possibility of achieving efficient light emission from Si nanostructures resulting in industrially viable monolithically integrated Si based optoelectronic devices and systems. Visible emissions from Si NCs in the range of red to blue have been observed depending on the particle size which has been attributed to both quantum confinement and surface passivation effects [3]. Surface oxidation also affects the luminescence [3, 4]. In the present study, we have synthesized Si NCs in bulk Si using a two stage heavy ion implantation and irradiation technique. In this, a low energy Au implantation has been used for amorphization of a top surface layer in a Si matrix of thickness ~ 30 nm. Si NCs are formed in this amorphised layer through a MeV Au ion induced localized crystallization. As the NCs are embedded in bulk Si, we do not expect any surface oxidation which is an advantage. The initial low energy Au implantation for amorphisation also introduces a significant amount of Au which is also known to form NCs dispersed in the top amorphous Si (a-Si) matrix. This has been found to be very important regarding enhancement in luminescence of the Si NCs produced [5]. In the present study Raman scattering has been used for characterization of the Si NCs.

EXPERIMENT

In the present study, Si (100) samples (n-type, 1-20 Ω.cm) were implanted with 32 keV Au$^-$ ions to fluence varying between 1×10^{15} - 1×10^{17} cm^{-2} to amorphize a top layer ~ 30 nm in thickness. These samples with a top amorphous layer were further irradiated with 1.5 MeV Au^{2+}

ions to a fixed fluence of 1×10^{15} cm^{-2}. All the implantations and irradiations were carried out at room temperature in the Ion Beam Laboratory at the Institute of Physics, Bhubaneswar. Some of the sequentially implanted samples were annealed at temperatures between 500° to 950°C for 1 hr in a quartz tube furnace. Raman scattering measurements were carried out on all the samples, using a RAMANOR U1000 Jobin-Yvon micro-Raman spectrometer equipped with a liquid nitrogen cooled charge-coupled device as the detector. An Ar-ion laser tuned to 514.5 nm was used as the excitation source. Raman spectra were recorded in the wavenumber range of 460-560 cm^{-1} in the backscattering geometry. One of the samples was subjected to high resolution transmission electron microscopy (TEM) for a direct look at the buried NCs. The results are presented in the next section. Since the MeV ion irradiation fluence, in all cases, was fixed, hereafter we shall refer the samples in terms of the fluence of the low energy implantation as carried out in the first stage. The samples initially implanted with 32 keV Au$^-$ ions to fluence up to 5×10^{15} cm^{-2} will be referred to as *Set I* while the same for fluence $\geq 1 \times 10^{16}$ cm^{-2} will be referred to as *Set II* samples.

DISCUSSION

In case of crystalline Si (c-Si), the first-order Raman scattering probes the optical phonon frequency at the Γ-point in the Brillouin-zone due to the $\Delta k = 0$ selection rule. This leaves a Raman active mode at 521 cm^{-1}, which gives a single line with natural line-width of 3.5 cm^{-1}. At nano-scale, this selection rule is relaxed. There is a softening and broadening of the first-order phonon mode in Raman scattering. This results in a shift of the Raman line towards lower wavenumber along with an asymmetric broadening. Compared to this, in amorphous Si (a-Si) the q-selection rule does not apply due to loss of long range order. In such a case all the phonons are optically allowed and the Raman scattering results in a broad hump at 480 cm^{-1} [6]. It is also very important to mention that structural damage, strain, alloying etc can also result in a downshift and broadening in Raman spectrum and it is important to identify the exact effect.

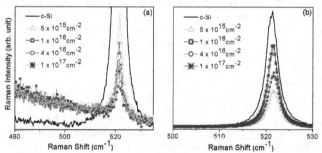

Figure 1: Raman scattering spectra of low energy Au implanted samples at various ion fluence. (a) Results on as-implanted samples, (b) Results after an annealing at 500 °C for 1 hr.

Typical Raman scattering spectra recorded after the initial low energy Au implantation, for various ion fluence, are shown in Fig. 1(a). A spectrum for an unimplanted c-Si sample is also included for comparison. For the implanted samples, one can clearly see a 521 cm^{-1} line along with a tail rising towards lower wavenumber. The intensity of the 521 cm^{-1} line is seen to drop with increase in Au fluence. One must note that the damaged / amorphised layer thickness

92

is about 30 nm which is smaller than the optical penetration depth of the exciting laser radiation which is about 100 nm in a-Si. The observed 521 cm^{-1} line is therefore due to the c-Si matrix lying beneath the top amorphised layer. The drop in intensity of this line is expected to be a result of increase in Au concentration and increased damage production (amorphisation) in the matrix which also results in the rising tail. In Fig 1 (b) we show the results on the same samples after an annealing at 500°C for 1 hr. One can see in all the implanted samples there is no rising tail below 520 cm^{-1} indicating almost complete crystallization of the amorphised region. The absence of any structure in this region also indicates the absence of Si NCs in the matrix. However, there are Au clusters in the system because of which the laser penetration depth has decreases resulting in reduced height of the Raman peak coming from c-Si lying underneath.

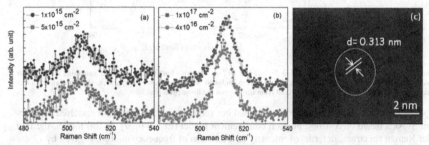

Figure 2: Background subtracted Raman scattering spectra corresponding to (a) Set- I and (b) Set- II samples. (c) A high resolution cross sectional TEM image taken on a sample with a Au fluence of 1x10^{15} cm^{-2}.

A subsequent 1.5 MeV Au irradiation is found to result in a Raman line below 521 cm^{-1}. This line is seen to ride over a large background coming from the amorphised layer. In order to see the Raman spectrum clearly it is necessary to subtract the background. Such data for the present samples are shown in Fig. 2 after subtraction of a linear background (between 480 and 540 cm^{-1}). The degree of downward shift in wavenumber and broadening in case of *Set- I* sample (Fig. 2 (a)) is seen to be more than that in *Set II* samples (Fig. 2 (b)). For *Set I* samples with low Au fluence the Raman peak appearing near 505 cm^{-1} is found to be rather weak and broad. On the other hand the same for *Set II* samples, appearing near 510 cm^{-1}, is seen to be quite strong. Further with increase in low energy Au implantation fluence there is a shift towards higher wavenumber which can be seen in *Set II* data. This indicates a growth in size of the Si NCs due to an overall higher energy deposition by the MeV Au beam. All the above samples were then annealed at 500° C for 1 hr. Interestingly no significant shift or change in intensity in the Raman peak could be observed for *Set I* samples. A representative cross sectional TEM image for an annealed sample with a low energy Au fluence of 5x10^{15} cm^{-2} is shown in Fig. 2 (c). In this figure one can clearly see nanocrystalline regions in an amorphised medium. The separation between the planes is found to be 0.313 nm corresponding to (111) planes of Si. Unlike the Set I sample, in case of *Set II* samples, the Raman peak shifted towards 521 cm^{-1} without much change in the linewidth. Some typical spectra corresponding to a low energy Au fluence of 4x10^{16} cm^{-2}, are shown in Fig 3 (a). As will be shown below, the observed downward shift in wavenumber following the annealing is due to removal of strain and defects in the NCs.

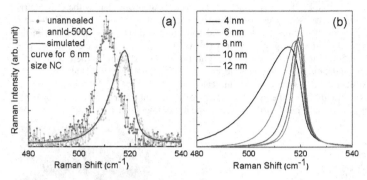

Figure 3: (a) Raman spectra of an initial high fluence [4 x 10¹⁶ cm⁻²] sample both before and after annealing at 500° C for 1 hr. A simulated spectrum for 6 nm size Si NC is also included in the figure for comparison. (b) Simulated Raman spectra for Si NCs of different sizes as obtained from the phonon confinement model.

To understand the data, we have carried out some estimation of Raman scattered intensity from Si NCs based on a strong phonon confinement model (PCM) [6, 7]. In this model, the first order Raman spectrum, in terms of intensity as a function of frequency ω, $I(\omega)$, is given by

$$I(\omega) = \int_0^1 \frac{\exp{(-q^2 L^2/4a^2)}}{[\omega - \omega(q)]^2 + (\Gamma_o/2)^2}\, d^3 q, \tag{1}$$

where q is expressed in units of $2\pi/a$, a being the lattice constant. The parameter L stands for the average size (diameter) of the NCs. Γ_o is the line-width of the longitudinal optical (LO) phonon in bulk c-Si (~ 3.6 cm⁻¹). The dispersion $\omega(q)$ of the LO phonon in c-Si is given by a parameterization of the calculated data as given by Tubino $et\ al.$ [8]

$$\omega^2(q) = A + B \cos{(\pi q/2)} \tag{2}$$

where $A = 1.714 \times 10^5$ cm⁻² and $B = 1.000 \times 10^5$ cm⁻². Using Eqns. 1 and 2, one can estimate the Raman scattered intensity spectrum for Si NCs of different diameters through a variation of L. Some results of this calculation for Si NCs of various sizes are shown in Fig. 3 (b). One can see a decrease in size of the NCs, results in both a broadening and an asymmetry in the Raman peak in addition to a shift towards lower wavenumber. It must be mentioned that within the framework of the PCM it was difficult explain the Raman peak as shown in Fig. 2 (b), based on a given size distribution. However, the peak structure obtained after annealing could be explained nicely using the data for a given size. This is clearly shown in Fig. 3 (a) where the results following the annealing at 500° C are seen to exactly match that obtained for a NC size of 6 nm. Based on the above findings we believe that the downward shift of the peak before annealing is mainly due to the presence of tensile strain which is known to compete with quantum confinement effect. The annealing results in the release of this strain in addition to removing some defects. Raman scattering spectra of few more $Set\ II$ samples, annealed at 560° and 650° C, for the same initial Au implantation fluence of 4x10¹⁶ cm⁻² are shown in Fig 4 (a). The same for annealing at 750°,

850° and 950° C are shown in Fig. 4(b) together with data for c-Si. Annealing at higher temperature is seen to result in a stronger and narrower Raman peak corresponding to larger NCs as is evident from a simulated data included in the figure. Annealing at 650° C is seen to result in NCs of average size 9 nm. Almost complete crystallization is seen to be achieved after annealing at 950° C.

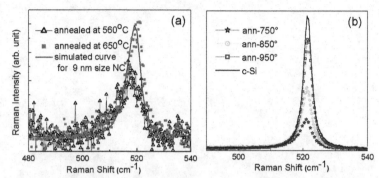

Figure 4:(a) Raman spectra of a initial high fluence [4 x 10^{16} cm^{-2}] samples annealed at 560° C and 650° C for 1 hr. A simulated spectrum for 9 nm size NC is also included in the figure. (b) Raman spectra obtained on samples annealed at 750°, 850° and 950° C for 1 hr. A spectrum corresponding to a c-Si sample is also included in the figures for comparison.

CONCLUSIONS

A two-stage Au implantation and irradiation technique has been used to synthesize Si NCs embedded in an a-Si matrix. Initially Si samples were implanted with 32 keV Au ions to amorphize the top surface. A subsequent 1.5 MeV Au irradiation has been found to result in formation of strained Si NCs in the top a-Si layer through ion beam induced localized crystallization. The observed tensile strain is seen to be removed following an annealing at 500° C for an hour. Annealing at higher temperature is found to result in a growth is NC size. Almost complete recrystallization has been seen following annealing at 950° C. Most importantly, Raman scattering results following an initial annealing indicate a very narrow size distribution of the NCs formed in the matrix. This technique which results in production of Si NCs with δ-function like size distribution is expected to be very useful for future applications.

REFERENCES

1. L. T. Canham, *Appl. Phys. Lett.,* **57**, 1046 (1990).
2. M.S. Hybertsen, *Phys. Rev. Lett.* **72**, 1514 (1994).
3. M. V. Wolkin, J. Jorne, P. M. Fauchet, G. Allan and C. Delerue, *Phys. Rev. Lett.* **82**, 197 (1999).
4. A. Puzder, A. J. Williamson, J. C. Grossman and G. Galli, *Phys. Rev. Lett.* **88**, 097401 (2002).

5. G. Sahu, H. Lenka, D. P. Mahapatra, B. Rout and F. D. McDaniel, *J. Phys. Cond. Matt*. (FTC) **22** 072203 (2010).
6. H. Ritcher, Z. P. Wang and L.Ley, *Solid State Comm.,* **39**, 625 (1981).
7. I. H. Campbell and P.M. Fauchet, *Solid State Comm.,* **58**, 739 (1986).
8. R. Tubino, L. Piseri and G. Zerbi, *J. of Chem. Phys.*, **56**, 1022 (1972).

Mater. Res. Soc. Symp. Proc. Vol. 1354 © 2011 Materials Research Society
DOI: 10.1557/opl.2011.1279

Effects of Hydrogen Ion Implantation and Thermal Annealing on Structural and Optical Properties of Single-crystal Sapphire.

William T. Spratt[1], Mengbing Huang[1,*], Chuanlei Jia[2], Lei Wang[3], Vimal K. Kamineni[1], Alain C. Diebold[1], Richard Matyi[1], and Hua Xia[4]

[1]College of Nanoscale Science and Engineering, University at Albany-SUNY, Albany, NY 12203, U.S.A.

[2] College of Physics Science and Technology, China University of Petroleum, Dongying, Shandong 257061, P. R. China.

[3]School of Physics and Microelectronics, Shandong University, Jinan, Shandong 250100, P. R. China

[4]RF and Photonics Laboratory, General Electric Global Research Center, Niskayuna, NY 12309, U.S.A.

* Email: mhuang@uamail.albany.edu

ABSTRACT

Due to its outstanding thermal and chemical stability, single-crystal sapphire is a crucial material for high-temperature optical sensing applications. The potential for using hydrogen ion implantation to fabricate stable, high temperature optical waveguides in single crystal sapphire is investigated in this work. Hydrogen ions were implanted in c-plane sapphire with energies of 35 keV and 1 MeV and fluences 10^{16}-10^{17}/cm^2. Subsequent annealing was carried out in air at temperatures ranging from 500°C to 1200°C. Complementary techniques were used to characterize the samples, including ellipsometry and prism coupling to examine optical properties, Rutherford backscattering/ion channeling for crystal defects, and nuclear reaction analysis for hydrogen profiling. Several guiding modes were observed in H-implanted (1 MeV) samples annealed above 800°C through prism coupling, and a maximum index modification of 3% was observed in the 35 keV samples and 1% in the 1 MeV samples through ellipsometry, with the 1 MeV index variation being confirmed through prism coupling. The possible causes of the index modifications, such as H related defects, as well as implications for tailoring the refractive index of sapphire are discussed.

INTRODUCTION

Optical waveguides are critical components of many diverse applications in integrated optics, ranging from optical computation, to photonic power delivery, to optical sensing. Increasingly these components are being used in harsh environments, supplanting existing technologies. Many methods exist for fabricating waveguides, of which ion implantation is an established technique for modifying the optical properties of materials. It allows a high degree of control over the location and extent of the modification while leaving other areas unmodified, which makes it an advantageous method for the fabrication of optical waveguiding structures in many materials[1,2].

Of particular interest are ionic crystals, such as sapphire (Al_2O_3), which have already been modified to serve as low loss optical and optoelectronic structures[3]. In addition to promising optical properties, sapphire possessed excellent thermal and chemical properties.

Studies have shown that the implantation of lighter ions in sapphire, such as H and He, induce a negative index change, and that heavier ions, such as O or Si, a positive one [1,4]. While low loss H implanted Al_2O_3 waveguides have already been demonstrated in other studies[5], they have focused mainly on the creation of optoelectronic structures, such as laser cavities, not the stability of the index layer at extreme temperatures [2,5]. In the present study we focus on H implanted Al_2O_3 and examine the dependence of the optical properties on annealing temperature. In addition we attempt to shed light on the origins of these changes, through the use of ion channeling (RBS/C) and hydrogen profiling (NRA).

EXPERIMENT

Two sets of sapphire samples were prepared. The first was a low energy implant of 35 keV which placed the implant at approximately 300nm below the surface, and the second was a high energy dual implant of 0.95 and 1 MeV to provide a wider modified layer around 8um below the surface which would allow waveguiding. Both implants we performed on c-plane sapphire wafers, with the fluences of the low and high energy implant being 6×10^{16} and 1×10^{17} ions/cm² respectively. The fluences were chosen to provide a similar defect density in both implants, so the analyses of both sample sets could be more directly relatable.

All implantations were performed at the Ion Beam Laboratory of the University at Albany on a 4 MeV Dynamitron accelerator at room temperature, using a water cooled sample holder. The wafers were then diced and annealed for 30 minutes in ambient atmosphere at temperatures up to 1100°C in a tube furnace, and to 1200°C in a box furnace. Channeling and hydrogen profiling measurements utilized the same accelerator, with the channeling using RBS at 3.135MeV and the profiling through Nuclear Reaction Analysis and the $^{15}N(p, \alpha \gamma)^{12}C$ resonant reaction at 6.385MeV. The RBS energy was chosen for oxygen resonance, taking into account energy loss before the beam reaches the implant depth.

Optical data was collected through prism coupling and ellipsometry. The high energy implanted samples were evaluated for waveguiding ability on a Metricon prism coupling system using TE polarized light at 632.8 nm. Index profiles were then calculated through the Reflectivity Calculation Method (RCM) using the prism coupling data [6]. Spectroscopic ellipsometric data was collected on a J.A. Woollam RC2 ellipsometer at wavelengths from 270 to 1700 nm, though in the high energy samples relevant data were obtained only above 1100 nm.

DISCUSSION

A few representative datasets from the prism coupling and ellipsometric measurements can be seen in Figure 1 (a) and (b), respectively. In the prism coupling experiments, seven waveguiding modes were observed, as indicated by the sharp dips in the spectra of Figure 1 (a). However, these modes were only observed in the samples annealed at or above 1000°C. The ellipsometry also does not show any significant variation until 1000 °C when some oscillations appear above 1300nm. Although ellipsometry does not show any oscillations below 1300nm for the high energy samples, this is most likely due to the increased absorption of shorter wavelengths in sapphire and the great depth. Previous work has shown the existence of a waveguiding region after implant without annealing, however this was not observed in our samples, possibly due to accumulated damage in the guiding region from the greater implant fluence [7].

Figure 1 (c) shows the index profiles derived from both prism coupling and ellipsometry, as well as the SRIM damage simulation with which the data show reasonable agreement [8]. Though seemingly great, the difference in index for the ellipsometric and prism coupling data is primarily due to the measurements being conducted at different wavelengths, even so, both methods calculate the maximum change in refractive index to within half a percent of each other, 0.6% for ellipsometry and 0.9% for prism coupling.

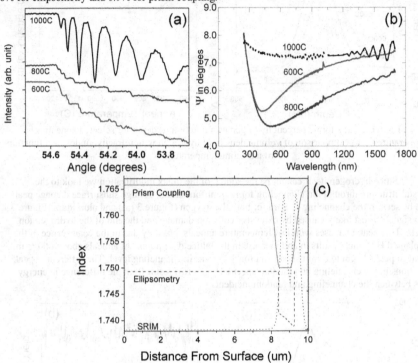

Figure 1. Example spectra from prism coupling (a) and ellipsometry (b). Both techniques are unable to detect significant change in index below 1000°C. (c) Comparison of prism coupling and ellipsometric data with expected damage profile from SRIM.

One potential cause of the index modification is the presence of a hydrogen inundated region within the sapphire directly causing a change in index. From figure 2 (a) however, we can see this is not the case: the atomic concentration of hydrogen is relatively stable at 4.5 at.% until above 1000°C, and has dropped to less than 0.5 at.% by 1100°C, therefore it cannot be the presence of hydrogen alone that is causing the change. The volume density of the hydrogen is shown in figure 2 (b), along the refractive index of the hydrogen layer for comparison where there appears to be a correlation between the hydrogen exodus at 1100°C and an index jump. It is interesting to note that in the high energy samples that support waveguiding, the hydrogen is no longer present in any significant concentration, which would suggest that the source of the index change is in modification of the crystal structure, and not the presence of hydrogen.

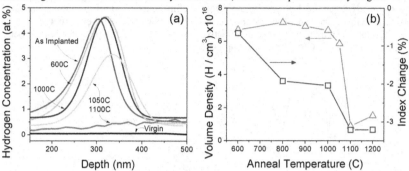

Figure 2. (a) Atomic percent hydrogen as a function of depth for selected annealing temperatures. (b) Comparison of volume density of hydrogen with refractive index as a function of annealing temperature.

Since hydrogen itself cannot be the source of the index modification we look to the crystal structure through ion channeling for more information. A large subsurface damage peak can be seen in the channeling spectra for both the oxygen (Figure 3 (a)) and aluminum (Figure 3 (b)) signals, most likely caused by extensive collision damage and the end of the hydrogen ion track. This peak increases with low temperature anneals, possibly due to the coalescence of the implanted H^+ which results in and increase in the lattice disruption. At anneals above 600°C the damage peaks begin to diminish and approach the virgin channeling level. The different O peak positions in the channeling and random spectra (Fig. 3 (a)) result from the difference in energy loss between the channeling and random incidence.

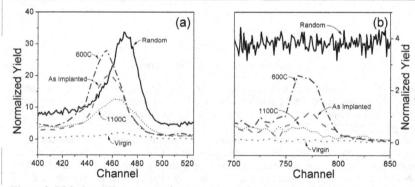

Figure 3. RBS channeling spectra of oxygen (a) and aluminum (b) of the low energy implant samples.

Ellipsometric measurements of these samples, which are shown alongside the χ_{min} values in Figure 4, indicate two large changes in refractive index, as well as a general decrease along with the χ_{min} values. The first large change occurs above 600°C, the temperature above which the χ_{min} decreases (after the initial increase), and the second above 1000°C, when the oxygen sublattice experience a jump in damage recovery. The χ_{min} values for aluminum and oxygen at 1200°C are 0.062 and 0.190, which are around twice the virgin levels of 0.016 and 0.083. The sudden decrease in the χ_{min} of the oxygen after 1100°C is most likely due to the defects in the oxygen sublattice requiring a higher anneal temperature to recover from the irradiation damage, which has been observed previously in sapphire [9].

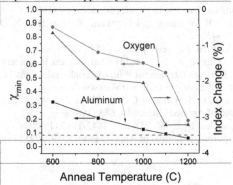

Figure 4. χ_{min} of the damaged regions and refractive index as a function of anneal temperature, the dotted lines indicate the virgin χ_{min}.

The waveguiding observed in the high energy implant samples may be caused by a buried defect layer created through implantation, possibly due to a decrease in density due to lattice expansion from damage at the end of the ion track or hydrogen clustering. The lack of observed waveguiding below 1000°C may be due to lattice damage in the guiding region, which is more easily repaired than the defects produced in the modified layer. Previous studies on hydrogen modification of sapphire have shown H bubble formation at temperatures less than 650°C, and an increase in bubble size of up to 70 nm at 1000°C, which would correlate with the damage trend in Figure 4 [10]. In addition, volume expansion has also been observed previously in sapphire, which would indicate a density related refractive index change [11].

CONCLUSIONS

Waveguiding structures which persist after high temperature annealing have been fabricated through ion implantation of hydrogen in sapphire. These structures build on previous work on sapphire waveguides and material analysis, by investigating the effects of high temperature annealing on waveguide formation and stability, as well as suggesting a cause of the index change. The refractive index modification appears to be independent of hydrogen concentration and due to their high temperature stability seem to offer the potential for extreme temperature waveguiding and sensing.

ACKNOWLEDGMENTS
This work has been supported by the National Science Foundation.

REFERENCES

1. F Chen, X Wang, and K Wang, Optical Materials **29**, 1523-1542 (2007).
2. P. D. Townsend, P. J. Chandler, and L. Zhang, Optical Effects of Ion Implantation (Cambridge University Press, Cambridge, 1994) p. 7.
3. L Laversenne, P Hoffmann, and M Pollnau, Applied Physics Letters 85, 5167-5169 (2004).
4. P. D. Townsend, Reports On Progress In Physics 50, 501–558 (1987).
5. C. Grivas, D. P. Shepherd, R. W. Eason, L. Laversenne, P. Moretti, C. N. Borca, and M. Pollnau, Optics Letters 31, 3450 (2006).
6. J. M. White and P. F. Heidrich, Applied Optics 15, 151 (1976).
7. N Sasajima, T Matsui, S Furuno, K Hojou, and H Otsu, Nuclear Instruments and Methods In Physics Research Section B: Beam Interactions With Materials and Atoms 148, 745-751 (1999).
8. Ziegler J F and Biersack J P computer code SRIM (www.srim.org).
9. H. Naramoto, C. W. White, J. M. Williams, C. J. McHargue, O. W. Holland, M. M. Abraham, and B. R. Appleton, Journal of Applied Physics 54, 683 (1983).
10. Y Katano, Journal of Nuclear Materials 258-263, 1842-1847 (1998).
11. G. B. Krefft and E. P. EerNisse, Journal of Applied Physics 49, 2725 (1978).

Mater. Res. Soc. Symp. Proc. Vol. 1354 © 2011 Materials Research Society
DOI: 10.1557/opl.2011.1214

Post-CMOS Integration of Nanomechanical Devices by Direct Ion Beam Irradiation of Silicon

Francesc Pérez-Murano[1], G. Rius[2], J. Llobet[1] and X. Borrisé[3]

[1]Institut de Microelectrònica de Barcelona (IMB-CNM, CSIC). Campus de la UAB, 08193 Bellaterra. Spain.

[2]Surface Science Laboratory. Toyota Technological Institute (TTI), 2-12-1 Hisakata, 468-8511 Nagoya. Japan

[3]Institut Català de Nanotecnologia (ICN). Campus de la UAB, 08193 Bellaterra. Spain

ABSTRACT

We present the development of CMOS compatible focused ion beam (FIB)-based method for the fabrication of nanomechanical devices. With only two step process, (i) patterning by direct exposure of silicon by the gallium beam and (ii) transfer of features to the structural layer by standard microfabrication silicon etching processes, operational devices are obtained. The ion beam modified silicon, acting as the etching mask, presents an outstanding robustness for both chemical and reactive ion etching process, enabling a simplified fabrication of nanomechanical devices with sub-micron resolution. As an example, single and double clamped silicon beams have been successfully produced. The compatibility check to guarantee the integrity of the electronic performance of CMOS circuits after the energetic beam irradiation is also investigated. Patterning based on direct ion beam exposure of silicon and etching presents advantages in comparison with more conventional lithography methods, such as electron beam lithography, since it is realized without the use of any resist media, which is especially challenging for the non-flat CMOS pre-fabricated substrates.

INTRODUCTION

Nanomechanical structures and nanoelectromechanical systems (NEMS) have a high potential to provide solutions for improving the performance of miniaturized systems in telecommunication, sensing or energy saving.

A nanomechanical device is a structure whose function is based on exploiting its mechanical properties (elasticity, resonance frequency, quality factor) [1]. Example of functional nanomechanical structures are cantilevered and, double clamped beams or nanowires. In order to provide a determined function (actuation, transduction, etc), the operation of a nanomechanical structure can be static or dynamic. In static mode, the deflection of the structure as a function of external or internal forces is used as the relevant magnitude. In dynamic mode, the nanomechanical structure is actuated at one or several of its resonance frequencies.

Nanomechanical structures operated in dynamic mode are used to build extremely sensitive mass sensors [2-6]. When a small quantity of mass is loaded on a nanomechanical structure, its resonance frequency changes. Monitoring the change of resonance frequency, the increase or decrease of mass can be monitored. The smaller the dimensions of the structure are, the more sensitive the device results. However, when the dimensions of the devices are reduced to submicron dimensions, experimental determination of the response of the device becomes challenging, being difficult to detect the change of mechanical characteristics. . In particular, optical methods are then difficult to implement since they require complex experimental set-up not useful in certain environments.

An alternative is to perform an all electrical actuation and read-out of the cantilever oscillation [7]. The simplest approach from a device fabrication point of view is based on detecting the current generated by the change of capacitance between the resonating structure and an electrode located in close proximity. However, when using capacitive detection in very small structures, the appearance of parasitic capacitances due to external connections, pads and electrical lines may generate additional currents that hide the desired signal.

In order to minimize undesired signal, a solution is to fabricate the nanomechanical structures monolithically incorporated into an integrated circuit [8]. In this way, the size of the electrical connections is minimized and the circuit amplifies the signal, while matching the impedance for the external equipment.

Defining nanometer-size structures incorporated into CMOS circuit requires the use of proper nanopatterning methods. Several approaches have been proposed, either i) using DUV lithography (in this case, the nanomechanical structure is fabricated simultaneously to the fabrication of the CMOS circuit, using the same materials and processes [9]), or ii) including extra post-CMOS processing methods based on electron beam lithography [10] or nanostencil lithography [11]. Here, we present a novel alternative nanopatterning method based on ion beam exposure of silicon. The main advantage is the simplicity of the process and the flexibility to be adapted to pre-patterned substrates.

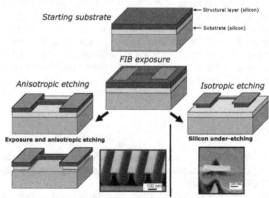

Figure 1. Overall process of fabricating nanomechanical structures by ion beam exposure and silicon etching. Left: ion beam exposure combined with anisotropic etching and release. Right: ion beam exposure and silicon etching.

EXPERIMENT AND RESULTS

The patterning method is based on local modification of the silicon surface and sub-surface using a focused ion beam. When directly exposing silicon to Ga$^+$ beam, two simultaneous processes occur: silicon removal and ion implantation. The occurrence of both processes produces that the final implantation profile (roughly 30 nm for a 30 keV Ga$^+$ beam [12]) is not dependent on the dose, which provides a robust process. The physical characterization of the material reveals that the exposed volume becomes highly amorphous silicon, with a 5% of gallium content, and without traces of oxide.

The key point to use this material for creating nanomechanical structures relies on the fact that irradiated features are highly resistant to silicon etching(see figure 1) . It has been shown [13-15] that the patterns defined by the selective ion beam exposure can be successfully transferred to the silicon by either anisotropic reactive ion etching or wet chemical etching of silicon. In addition, isotropic silicon etching can be used, to under-etch the silicon below the modified volume. In this case, the exposed material becomes the ultrathin structural layer of the nanomechanical structure. The use of substrates silicon on insulator (SOI) type allows that the structures remain electrically insulated from the surrounding.

Figure 2 show some examples of structures fabricated using ion beam exposure and silicon etching. We use a typical exposure dose of $6.2 \cdot 10^{16}$ ions/cm^2. As it has been mentioned above, the depth of the implanted volume is around 30 nm, what determines the thickness of the structure, while the lateral dimensions are mainly set by the beam diameter on the surface and lateral spread of ions in silicon.. Figure 2 shows some examples of structures fabricated by ion beam exposure and TMAH etching: a cantilever of 5 μm length and 150 nm width (figure 2.a) an array of nanowires with a diameter around to 30 nm (figure 2.b), and an example of coupled nanomechanical structures, where the two nanowires are connected by a thin membrane. Figure 2.d shows an example of how taking advantage that the implanted profile is independent of the quantity of silicon milled, complex three-dimensional nanostructures can be generated..

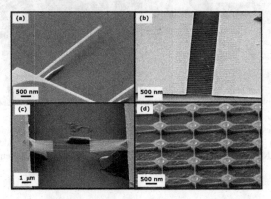

Figure 2. Examples of nanomechanical structures fabricated by ion beam exposure and anisotropic silicon etching.

Electrical actuation and capacitive detection of the devices requires that the structures to be conductive. Figure 3.a shows the process to fabricate nanomechanical resonators with electrical contacts. First, metallic microelectrodes are fabricated on a SOI substrate by conventional optical lithography, metal deposition and resist lift-off. We have chosen tungsten as the material for the electrode because it can sustain a high temperature annealing process: as it has been said before, the exposed material is highly amorphous, which may degrade its mechanical properties. By submitting the material to a medium or high temperature annealing, crystallinity is recovered, which would improve the performance. Figures 3.b and c show SEM example of nanomechanical devices fabricated with electrical contacts. We have used these structures to check the electrical resistivity of the exposed material. The value obtained is around 4.5 Ω·cm, similar to what it is found for amorphous silicon.

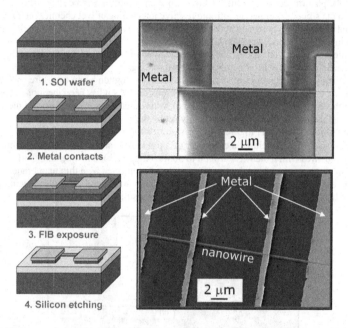

Figure 3. (a) Process to create nanomechanical structures with electrical contacts. (b) Example of nanowire with electrical contacts and an additional metallic electrode for capacitive measurements. (c) Nanowire with metal lines for four-probe resistivity measurements

DISCUSSION

The main advantages of ion beam exposure plus silicon etching as fabrication process of nanomechanical structures lies on its remarkable simplicity, for example, as compared to electron beam lithography, where the number of process steps and complexity increase (electron beam lithography is based on exposing a resist that has to be deposited by spinning, exposed, revealed and eliminated after the silicon etching). But, in addition, as compared to electron beam irradiation, the potential damage to charge sensitive substrates is decreased. The reason for this is that the penetration range for energetic ion beams in solids is far shorter than for electrons. It is known that an electron beam can damage integrated circuits, since it can create charged defects in the gate oxide. By using properly defined test structures, we have shown previously [16] that ion beam exposure of silicon is fully compatible with CMOS substrates as far as the ion beam is not directly and intensively delivered on top of the CMOS circuit.

In terms of patterning speed, ion beam exposure using focused ion beams, although slightly faster than electron beam lithography, is by far much slower than DUV lithography, since it is a serial beam-based process. However, there is the possibility to improve its speed if it is intended for future industrial applications, where massive manufacturing is required. A first option would be to use systems based on multiple ion beams, as those recently developed [17]. A second option is to define re-usable hard mask (stencils masks) and combine it with broad beam ion implantation.

CONCLUSIONS

Incorporation of nanomechanical devices into integrated nanomechanical circuits requires the use of advanced patterning methods. The process based on ion beam exposure of silicon and TMAH etching is very convenient because of its simplicity, robustness and CMOS compatibility. However, further use of the process in view of future industrial applications have the drawback of low throughput due to its serial beam-based nature when using focused ion beams. Alternatives like the use of parallel e-beam systems or broad beam ion implantation requires gaining knowledge about the mechanism why the exposed volume of material becomes strongly resistant to silicon etching. At present this chemical property is not fully understood for the case of gallium from a focused ion beam system, and it is unknown if a similar phenomena can be exploited for other kind of ion beams and substrates.

ACKNOWLEDGMENTS

This work has been partially supported by European project CHARPAN, NMP2-CT-2005-515803, and Spanish project, NANOFUN, TEC2009-14517-C02-01.

REFERENCES

1. A. N. Cleland. *Foundation of Nanomechanics. From solid state devices to applications.* Springer Verlag, 2003.
2. K.L. Ekinci, Y. T. Yang, M. L. Roukes. *J. Appl. Phys.* **95**, 2682 (2004)
3. J. Verd, A. Uranga, G. Abadal, J. Teva, F. Torres, F. Pérez-Murano, J. Fraxedas. *Appl. Phys. Lett.* **91**, 013501 (2007)
4. J. Arcamone, M. Sansa, J. Verd, A. Uranga, G. Abadal, N. Barniol, M. van den Boogaart, J. Brugger, F. Perez-Murano. *Small*, **5**, 176-180 (2009)
5. B. Lassagne, D. Garcia-Sanchez, A. Aguasca, A. Bachtold. Nanoletters 8, 3735 (2008)
6. K. Jensen, K. Kim, A. Zettl, Nature *Nanotechnology* **3**, 5333 (2008)
7. G. Abadal, Z. J. Davis, B. Helbo, X. Borrisé, R. Ruiz, A. Boisen, F. Campabadal, J. Esteve, E. Figueras, F. Pérez-Murano, N. Barniol. *Nanotechnology*, **12**, 100 (2001)
8. Z. J. Davis, G. Abadal, B. Helbo, O. Hansen, F. Campabadal, F. Pérez-Murano, J. Esteve, E. Figueras, J. Verd, N. Barniol, A. Boisen. *Sensors and Actuators A* **105**, 311-319(2003)
9. J.Verd, G.Abadal, M.Villarroya, J.Teva, A.Uranga, F.Campabadal, J.Esteve, E.Figueras, F.Pérez-Murano, Z.Davis, E.Forsen, A.Boisen, N.Barniol. *Jl of Microelectromechanical Systems*, **14**, 508-519 (2005)
10. E. Forsén,G. Abadal, S. Ghatnekar-Nilsson, J. Teva, J. Verd, R. Sandberg, W. Svendsen, F. Pérez-Murano, J. Esteve, E. Figueras, F. Campabadal, L. Montelius, N. Barniol, and A. Boisen. *Applied Physics Letters*, **87**, 04357, (2005)
11. J. Arcamone, M. A. F. van den Boogaart, F. Serra-Graells, J. Fraxedas, J. Brugger and F. Pérez-Murano. Nanotechnology **19**, 305302 (2008)
12. L.A. Gianuzzo and F.A. Stevie, *Ed. Introduction to focused ion beams.* Springer. 2005
13. B. Schmidt, S. Oswald, L. Bischoff. *Journal of the Electrochemical Society*, **152**, G875 (2005)
14. N. Cherukov, K. Grigoras, A. Peltonen, S. Franssila, I. Tottonen. *Nanotechnology*, **20**, 065307 (2009)
15. G. Rius, J. Llobet, X. Borrisé, N. Mestres, A. Retolaza, S. Merino , F. Perez-Murano *J. Vac. Sci.Technol. B* **27**, 2691 (2009)
16. G. Rius, J. Llobet, M. J. Esplandiu, L. Solé, X. Borrisé, F. Perez-Murano. *Microelec.Eng.* **86**, 892 (2009)
17. C. Klein, E Platzgummer, H. Loeschner. *Proceedings of SPIE* **7545**, DOI: .1117/12.863143, (2010)

Mater. Res. Soc. Symp. Proc. Vol. 1354 © 2011 Materials Research Society
DOI: 10.1557/opl.2011.1280

Enhanced adhesion of coating layers by Ion Beam Mixing: An application for nuclear hydrogen production

Jae-Won Park, Hyung-Jin Kim, Sunmog Yeo and Seong-Duk Hong

Korea Atomic Energy Research Institute, Daejon-City, South.Korea

ABSTRACT

The bonding between two dissimilar materials has been a problem, partiularly in coating metals with non-metallic protective layer. In this work, it is demonstrated that a strong bonding between ceramics/metal can be achieved by mixing the atoms at the interface by ion-beam. Specifically, SiC coating on Hastelloy X was studied for a high temperature corrosion protection. Auger elemental mapping across the interface shows a far broader mixed region than the region expected by SRIM calculation, which is thought to be due to the thermal spike liquid state diffusion. The results showed that, although the thermal expansion coefficient of Hastelloy X is about three times higher than that of SiC, the film did not peel-off at above 900 °C confirming excellent adhesion. Instead, the SiC film was cracked along the grain boundary of the substrate above 700 °C. At above 900 °C, the film was crystallized forming islands on the substrate so that a considerable part of the substrate surface could be exposed to the corrosive environment. To cover the exposed area, it is suggested that the coating/IBM process should be repeated multiply.

INTRODUCTION

High Temperature Gas Cooled Reactor (HTGR) combined with the Iodine-Sulfur (IS) cycle has been regarded as the most efficient system for a mass production of hydrogen [1,2]. In the IS cycle, a process heat exchanger (PHE) comprised of channels for He and decomposed sulfuric acid gas ($SO_2/SO_3/H_2O$) is needed. The material used for the sulfuric acid gas channels is subjected to such a severe corrosion environment, however, there is no suitable commercial metallic material available presently. For this reason, consideration has been given to surface modification of metallic materials .

We selected Hastelloy X as the metallic substrate due to its good mechanical properties at a high temperatures and SiC as a corrosion inhibiting coating material (its corrosion resistance being due to the very strong covalent bonding between silicon and carbon [3,4]). This ceramic-coating-on-metal system certainly has considerable merit because that it does not hamper the manufacturabilty of the system as compared to a ceramic PHE system. The prime concern in such a combination is the adhesion between the film and the substrate at elevated temperatures due to the large difference in the thermal expansion coefficients. (CTE of Hastelloy X:$16.6x10^{-6}$ at 980°C and CTE of SiC :$5.0x10^{-6}$ at 1000°C).

In this paper, excellent bonding between e-beam deposited SiC film and Hastelloy X metal is demonstrated by employing ion mixing. An Auger elemental mapping data obtained across the interface of the SiC film-Hastelly X substrate are discussed. The surface morphology change as a function of temperature was observed with SEM and optical microscopy and the formation of new phases at the interfacial was analyzed with XRD.

EXPERIMENT

SiC films were deposited with an electron beam evaporative method on metallic substrate (Hastelloy X sheet: ~ 15x15x0.5mm) surface-polished by a diamond paste up to 0.5 μm. Prior to a SiC deposition, a sputter cleaning of the sample was carried out for 10 minutes with an N ion energy of ~10 keV and a current of 0.5 Ampere. Then, the electron beam evaporative deposition of the SiC was performed to 50 nm thickness, followed by a nitrogen ion beam mixing at 70keV with a dose of ~ 5x10^{16} ions/cm^2. A further SiC evaporative deposition up to a total of ~ 1 μm was then conducted with a deposition rate of ~3 Å/s produced by an electron beam current of ~0.15 A. The substrate temperature during the e-beam evaporative deposition was ~150 °C.

Elemental distributions at the interface before and after the ion beam bombardment were observed by Auger elemental mapping. Scanning Auger Microprobe (SAM) Phi model 670 was employed for the mapping and the elemental depth profiling. Before data acquisition with SAM, the surface of the cross-sectioned sample was gently sputtered with 500 eV Ar ions for 30 s to eliminate the surface contaminants. Cross-sectional samples were prepared and the Si KLL peak was chosen for the mapping. Auger peaks subtracting the baseline from the AES spectra at each analytical point, in which the contrast corresponds to the amount of the existing elements.

The coated/ion beam mixed samples were heated up to 950 °C and the surface morphologies were observed with SEM.

RESULTS AND DISCUSSION

According to the calculation using the SRIM (Stopping Range of Ions in Matter) code [5], the stopping range for 70 keV N ions is near 0.2μm (200 nm) depth for the low density as-deposited SiC film (1.92 g/cm^3) and near 0.13 μm depth for a bulk SiC (3.217 g/cm^3). Since the ion penetration depth (200 nm) is about four times larger than the thickness of deposited film (50nm), the ion mixing should occur across the film/substrate interface. N-ion bombardment makes the interface not only mix but also reinforce the metallic substrate by forming nitrides with implanted nitrogen [6 - 8].

Figure 1 shows Si mapping acquired by AES at the interface between the SiC film and the Hastelloy-X substrate before (Fig. 1a) and after (Fig. 1b) ion beam bombardment. The mapping in Fig. 1 is a result of the data acquired for a couple of days. White areas in each map show Si distribution. After implantation, the range of white area extended further into the dark metal area (Fig. 1b), revealing the Si penetration across the interface. The interface is indicated with a dotted line. Carbon mapping was not done, but carbon also is expected to be present in the metal region, possibly some forming SiC, because the recoil implantation should have occurred for both Si and C, although the recoil behaviors may differ for Si and C. The graded distribution of Si-intensity across the interface shows clearly that the intermixing took place at the interface. Since this analysis is based on the Si mapping, the degree of intermixing for other elements is unknown, although atoms in the film and the substrate are expected to mix into the opposing layers during ion mixing. As seen in Fig.1b, a far broader mixed region (~ 1μm) than expected by SRIM calculation(~0.15μm) is observed. For this, the thermal spike and liquid phase interdiffusion model that the collision cascade has temperatures exceeding the melting point may be considered as a dominant mechanism [9].

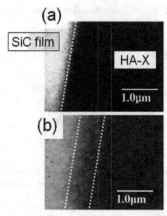

(a)

SiC film

HA-X

1.0µm

(b)

1.0µm

Figure 1. Auger elemental mapping across the interface shows a far broader mixed region than the region expected by SRIM calculation, which is thought to be due to the thermal spike liquid state diffusion.

Our previous XRD study showed several unidentified X-ray peaks in addition to the known peaks from fcc Hastelloy X and fcc SiC [10]. The presence of the crystalline SiC-peaks indicated the as-deposited amorphous film crystallized during annealing. Some of the peaks appeared to originate from SiO_2, surface oxide on the deposited film. However, there were many unidentified peaks. They are thought to come mainly from Ni-Si compounds such as Ni_2Si, Ni_2SiC, Ni_5Si_2, Cr-O, etc. Integrated intensity of unidentified peaks are stronger after annealing at 1000 °C than 900 °C, suggesting that the amount of the new phases increases by thermally activated diffusion process. The formation of Ni-Si compounds was also investigated for Ohmic contact of Ni on SiC for high temperature and high power devices [11 - 18]. Jacob et al. [11] observed a complete reaction between 5000 Å Ni film and SiC substrate at 900 °C. Cho et al. [12] also reported Ni_2Si formation at temperatures greater than 600 °C using a glancing angle XRD method, and that the Ni_2Si formation increased progressively with increasing temperatures from 700 °C to 900 °C, converting Ni to Ni_2Si almost completely at 900 °C. So far, no detailed study has been reported on the formation of Ni-Si compounds for a SiC-coated Ni base super alloy system.

Fig.2 is schematic description how the interfacial reaction takes place by ion beam mixing as the temperature increases. This drawing explains that the ion mixing plays a role in fastening a thin ceramic material to a metallic substrate until the interfacial chemical reaction takes place, while the coating without ion beam mixing may be easily detached before the interfacial reaction onset.

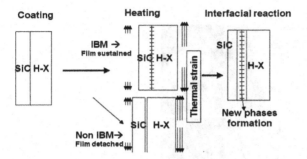

Figure 2. Schematics of interfacial reaction by ion beam mixing at the elevated temperature: The ion mixing can be used as a tool in fastening a thin ceramic material to a metallic substrate until the interfacial chemical reaction takes place.

In developing a coating for the nuclear hydrogen production application, it is essential to anneal the coated film at high temperatures above 900 °C to ensure the formation of the protective coating at the designed service temperature. However, high temperature annealing does not always warrant the formation of stable compounds at the interface region in any ceramic-metal system, unless the reaction between the two dissimilar materials is favored thermodynamically, that is, the mixed layer segregates back into its components as the temperature increases if the ΔH_{mix} are positive [19]. For instance, the ion mixed SiC film on the stainless steel substrate delaminates completely at 900 °C [20]. In view of the favorable thermodynamics, SiC-Hastelloy is considered a good candidate system for ion mixing in designing a protective coating used at high temperatures.

Although the strong adhesion is achieved, there still remains a problem of the film crack caused by the lower thermal expansion coefficient of the ceramics than metals. Fig. 3 shows SEM micrographs of

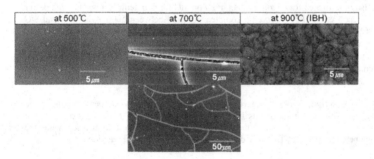

Figure 3. SEM micrographs of the film morphology of SiC film-Hastelloy X system as a function of temperature: At ~ 700 °C, the coating is cracked along the grain boundary of H-X substrate. At ~ 900 °C, the coated is crystallized forming the film islands.

the film morphology of SiC film-Hastelloy X system as a function of temperature. In spite of about three times higher thermal expansion coefficient of Hastelloy X than SiC at ~ 900 °C the film did not peel-off at above 900 °C confirming excellent adhesion. Instead, the SiC film was cracked at ~700 °C.

It looks the crack forms along the grain boundaries of the Hastelloy-X substrate. The grain boundaries on the surface of Hastelloy X substrate seems to have been developed by a thermal etching since we finished the substrate surface only with grinding/polishing before the film deposition, therefore, there should have been no grain boundary developed on the surface initially. At above 900°C, the film was crystallized forming islands on the substrate so that a considerable part of the substrate surface could be exposed to the corrosive environment.

To cover the exposed area, the coating/IBM process needs to be repeated multiply. The effects of the multiple processes will be reported in detail elsewhere.

CONCLUSIONS

A strong bonding between ceramics/metal can be achieved by mixing the atoms at the interface by ion-beam and forming a graded composition. Auger elemental mapping across the interface shows a far broader mixed region than the region expected by SRIM calculation, which is thought to be due to the thermal spike liquid state diffusion. The mechanism may be that the ion beam mixing fastens the SiC coated layer with the Hastelloy X substrate until the interfacial reaction occurs. Then, the new phase is formed at the film/substrate interface by interfacial reaction, which acts as a functionally degraded layer. Although the thermal expansion coefficient of Hastelloy X is about three times higher than that of SiC, the film did not peel-off at above 900 °C confirming excellent adhesion. Instead, the SiC film was cracked along the grain boundary of the substrate above 700°C. At above 900 °C, the film was crystallized forming islands on the substrate so that a considerable part of the substrate surface could be exposed to the corrosive environment. To cover the exposed area, it is suggested that the coating/IBM process should be repeated multiply.

ACKNOWLEDGMENTS

This study was supported by both Nuclear Hydrogen Development Project and Proton Engineering Frontier Project sponsored by Ministry of Education, Science and Technology, Republic of Korea.

REFERENCES

1. Hiroyuki Ota, Shinji Kubo, Masatoshi Hodotsuka, Takanari Inatomi, Masahiko Kobayashi, Atuhiko Terada, Seiji Kasahara, Ryutaro Hino, Kenji Ogura, Shigeki Maruyama, 13th International Conference on Nuclear Engineering, Beijing, China, May 16-20, 2005, ICONE-13-50494
2. S. Fujikawa et. al., J. Nucl. Sci. Technol., Vol. 41 (2004) 1245
3. J.-P. Riviere, J. Delafond, P. Misaelides, F. Noli, Surf. Coat. Technol. 100-101 (1998) 243
4. S. Fujikawa, H. Hatashi, T. Nakazawa, K. Kawasaki, T. Iyoku, S. Nakagawa and N. Sakaba., J. Nucl. Sci. Technol., 41 (2004) 1245
5. J.F. Ziegler, J.P. Biersack and U. Littmark, The Stopping and Range of Ions in Solids, Pergamon

Press, New York, 1985.

6. J. Tian, Q. Zhang, L. Xia, S. F. Yoon, J. Ahn, E. S. Byon, Q. Zhou, S. G. Wang, J. Q. Li, D. J. Yang, Mater. Res. Bulletin, 39(2004)917-922.

7. I. M. Neklyudov, A. N. Morozov, Physica B 350 (2004) 325-337..

8. K. Nakajima, S. Okamoto, T. Phada, Journal of Applied Physics 65(1989)4357.

9. M. Nastasi and J.W. Mayer, Materials Science and Engineering, R12 (1994) 1-52.

10. J. Park, Z. S. Khan, H. Kim, Y. Kim, Mater. Res. Symp. Proc., Vol. 1125 (2009)65

11. C. Jacob, P. Pirouz, H.-I. Kuo and M. Mehregany, Solid-State Electronics Vol. 42, No. 12 (1998) 2329-2334.

12. Cho, H. J., Hwang, C. S., Bang, W. and Kim, H. J., in Silicon Carbide and Related Materials, ed. M. G. Spencer, R. P. Devaty, J. A. Edmond, M. Asif Khan, R. Kaplan and M. Rahman. Inst. Phys. Conf. Ser. 137, Bristol and Philadelphia (1994) 663.

13. Rastegaeva, M. G., Andreev, A. N., Zelenin, V. V., A. I., Babanin, Nikitina, I. P., Chelnokov, V. E. and Rastegaev, V. P., in Silicon Carbide and Related Materials 1995, ed. S. Nakashima, H. Matsunami, S. Yoshida and H. Harima. Inst. Phys. Conf. Ser. 142, Bristol and Philadelphia, 1996, p. 581.

14. Liu, S., Reinhardt, K., Severt, C. and Sco®eld, J., in Silicon Carbide and Related Materials 1995, ed. S. Nakashima, H. Matsunami, S. Yoshida and H. Harima. Inst. Phys. Conf. Ser. 142, Bristol and Philadelphia, 1996, p. 589.

15. Steckl, A. J., Su, J. N., Yih, P. H., Yuan, C. and Li, J. P., in Silicon Carbide and Related Materials, ed. M. G. Spencer, R. P. Devaty, J. A. Edmond, M. Asif Khan, R. Kaplan and M. Rahman. Inst. Phys. Conf. Ser. 137, Bristol and Philadelphia, 1994, p. 653.

16. O.J. Guy , G. Pope, I. Blackwood, K.S. Teng, L. Chen, W.Y. Lee, S.P. Wilks, P.A. Mawby, Surface Science 573 (2004) 253–263.

17. S.Y. Han, J.-L. Lee, J. Electrochem. Soc. 149 (3) (2002) G189.

18. F. La Via, F. Roccaforte, A. Makhtari, V. Raineri, P.Musumeci, L. Calcagno, Microelectron. Eng. 60 (2002) 269–282.

19. R.S. Averback and D.Peak, Appl. Phys., A39 (1986) 59.

20. Jae-Won Park, Youngjin Chun, Jonghwa Chang, Journal of Nuclear Materials, Vol. 362, Issues 2-3 (2007) 268-273.

Mater. Res. Soc. Symp. Proc. Vol. 1354 © 2011 Materials Research Society
DOI: 10.1557/opl.2011.1215

Structural Changes Induced by Swift Heavy Ion Beams in tensile strained Al $_{(1-x)}$In$_x$N /GaN Hetero-structures

G. Devaraju, Anand P. Pathak, N. Srinivasa Rao, V. Saikiran, N. Sathish and S. V.S Nageswara Rao

School of Physics, University of Hyderabad, Hyderabad 500046, A P, India

ABSTRACT

We report here swift heavy ion (SHI) irradiation induced effects on structural and surface properties of III-nitrides. Tensile strained Al$_{(1-x)}$In$_x$N/GaN Hetero-Structures (HS) were realized using Metal Organic Chemical Vapour Despotion (MOCVD) technique with indium composition as 12%. Ion species and energies are chosen such that electronic energy deposition rates differ significantly in Al$_{(1-x)}$In$_x$N and are essential for understanding the ion beam interactions at the interfaces. Thus the samples were irradiated with 80 MeV Ni^{6+} and 100 MeV Ag^{7+} ions at varied fluence (1×10^{12} and 3×10^{12} ions/cm^2) to alter the structural properties. Under this energy regime, the structural changes in Al$_{(1-x)}$In$_x$N would occur due to the intense ultrafast excitations of electrons along the ion path. We employed different characterization techniques like High Resolution X- ray Diffraction (HRXRD) and Rutherford back scattering spectrometry (RBS) for composition, thickness and strain. HRXRD and RBS experimental spectra have been fitted with Philip's epitaxy SIMNRA code, which yields thickness and composition from compound semiconductors. The surface morphology of pristine and irradiated samples is studied and compared by Atomic Force Microscopy (AFM).

INTRODUCTION

III-Nitride compound semiconductors have tremendous applications in opto-electronic, high frequency and high power devices. In spite of huge defect densities [1] (four orders of magnitude higher than those in their counterpart compounds based on GaAs), these materials show excellent luminescence and electrical properties. Among family of nitrides, the Al$_{(1-x)}$ In$_x$N alloys are much less investigated than In$_x$Ga$_{(1-x)}$N and Al$_x$Ga$_{(1-x)}$N because it is very difficult to obtain high-quality Al$_{(1-x)}$ In$_x$N layers due to significant difference of thermal stability between InN and AlN [2–4]. However, Al$_{(1-x)}$ In$_x$N is a promising candidate in Bragg reflectors and field effect transistors[5-6]. Due to their technological importance in satellite communications, there is a need to understand how device performance degrades under radiation treatment of these semiconductor materials. Much has not been explored but it is observed that high-energy ions create carrier traps which compromise electrical and optical properties [7, 8]. Hence there is great demand to understand the interfaces of these HS, their strain and surface morphology under the effects of heavy ion irradiation.

In this work, the Al$_{(1-x)}$In$_x$N Hetero-Structures (HS) were grown on sapphire with GaN templates on c plane sapphire. These structures were irradiated with Ni and Ag ions.

Subsequently, the surface morphology was studied by Atomic Force Microscopy (AFM). It is found that ion irradiation can obviously influence the surface morphology. The (0002) symmetric ω-2θ scans are recorded and results are compared for pristine and irradiated samples. Consequently, strain from pristine and irradiated samples has been estimated and compared. Composition obtained from High Resolution X-ray diffraction (HRXRD) for pristine sample have been compared with those obtained with Rutherford back scattering spectrometry (RBS) results.

EXPERIMENTAL DETAILS

The studied AlInN/GaN HS were grown on (0001) sapphire by Metal Organic Chemical Vapour Deposition (MOCVD). Gas precursors were trimethyl-Al, trimethyl-In and ammonia (NH_3), using 50 mbar of pressure, and AlInN layer was deposited at 860 ˚C. Then samples were cut into 1×1 cm^2 and mounted on a Cu ladder with a conductive silver paste, and a low beam current was maintained for Ni and Ag ions during irradiation to avoid heating of samples. Subsequently, samples were oriented to 7° with respect to the beam axis to avoid any channeling of ions. Then samples were irradiated with 80 MeV Ni and 100 MeV Ag ions at different fluences (1×10^{12} and 3×10^{12} ions/cm^2) by scanning over 1×1 cm^2 area using 15 MV Pelletron accelerator at Inter University Accelerator Center (IUAC), New Delhi. Stopping and Ranges of Ions in Matter (SRIM) [9] code was used to calculate the electronic energy loss (S_e) and projected ranges for 80 MeV Ni and 100 MeV Ag ions. The corresponding average S_e and projected range of ions in AlInN for Ni ions are 10 keV/nm and 11 μm, and for Ag ions are 17 keV/nm and 10 μm, respectively. In this case, the projected ranges of ions are larger than the thickness of sample. Hence, these ions lose their energy predominantly in the electronic stopping power regime, while the end-of-range (EOR) region is deep inside the sapphire substrate. The structural properties of as deposited and subsequently irradiated $Al_{(1-x)}In_xN$ films were characterized by HRXRD. These measurements around (0002) reflection were performed using a Bruker D8 DISCOVER equipped with a Cu K_a ($\lambda = 0.154$ nm) radiation, a Bragg mirror and four-bounce Ge (220) monochromator. The diffracted X-rays were collected through Ge (022) analyzer crystal. RBS studies were performed with a 0.2×0.6 mm^2 collimated beam of 2 MeV He$^+$ ions. The backscattered particles were detected at 165° with respect to the incident beam direction using a silicon surface barrier detector with an energy resolution of 15 keV. SIMNRA code has been used to fit RBS spectrum collected from pristine sample. Surface morphology of samples was investigated by AFM (in Dynamic Force Microscopy mode) using SPA 400, Seiko Instruments Inc.

RESULTS AND DISCUSSION

As shown in Fig. 1, (0002) symmetric ω-2θ scan of pristine sample is carried out in triple axis geometry. Simulation of the HRXRD (Fig.1) has been carried out using the dynamical theory based Philips Epitaxy software [10]. Composition of the layer has been optimized by a trial and error method starting from some nominal values until a satisfactory fit is observed. The simulated scan matches reasonably well with the experimental one for nominal Indium composition (12%). Random and simulated spectra of $Al_{1-x}In_xN$/GaN HS from RBS experiments are shown in Fig 2. The elemental concentration was determined by means of the SIMNRA simulation [11]. The arrows labeled In, Al and Ga indicate the energy of the backscattered He$^+$

ion from In and Al atoms at the surface and Ga atoms from GaN layer. The In signal is completely separated from the Ga signal and similarly, Al signal is clearly distinguishable from Ga in the spectrum. The window (between 560 -600 channels) shows gradient of In and Ga as a result of inter diffusion of In and Ga from AlInN and GaN layers due to differences in ambient temperature during growth. This results in the formation of an intermediate thin layer of AlInGaN. Consequently the observed concentration of In is found to be 6 % in near surface region.

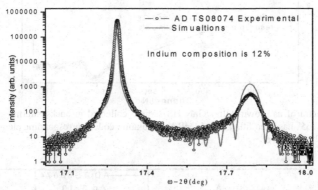

Fig.1) (0002) symmetric ω-2θ rocking curve of pristine AlInN/GaN HS fitted with Philip's epitaxy simulation

The Ni and Ag ions irradiated ω-2θ rocking curves exhibited significant changes in AlInN (0002) peak positions with respect to as deposited sample, as shown in Fig 3. Here the effects of electronic energy loss (S_e) and fluence on Bragg angle can also be seen. It is evident that Bragg angle for AlInN layer increases with increase of fluence due to enhancement in strain values. This is attributed to point defects created by heavy ions resulting in lattice expansions or compressions which in turn changed the strain values. Interestingly at lower fluence, Ni and Ag ions irradiation has resulted in same strain values. Higher S_e has further increased the strain as summarized in Table 1.

Table.1   Strain and RMS roughness of pristine and irradiated $Al_{(1-x)}In_xN$/GaN HS

Sample Details	Angular separation between GaN and AlInN Bragg peaks	RMS roughness from AFM
AD TS08074	0.49°	5 nm
80 MeV Ni ions @ 1×10^{12}	0.51°	5 nm
80 MeV Ni @ ions 3×10^{12}	0.53°	4.2 nm
100 MeV Ag ions @1×10^{12}	0.51°	4.2 nm
100 MeV Ag ions @ 3×10^{12}	0.56°	3.6 nm

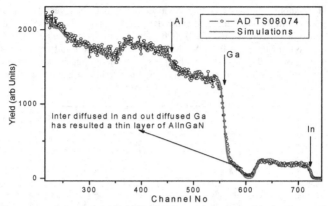

Fig 2) RBS spectra of as grown AlInN/GaN HS sample collected in backscattering geometry with detector at 165° has been fitted with SIMRA simulation code. Corresponding elements like In, Ga, Al and N are labeled.

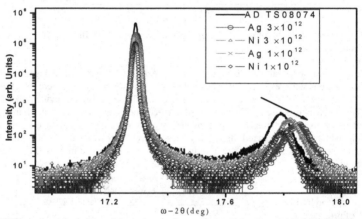

Fig 3: (0002) symmetric ω-2θ rocking curves of as grown and irradiated AlInN/GaN HS

Fig.4 shows typical AFM images with a 2 x2 μm^2 scan area taken from the pristine and irradiated tensile strained AlInN/GaN HS. It is known that grain size depends on growth conditions like gas flow rate, growth temperature etc. In the present study AFM images clearly show surface morphology with uniform grain sizes for the pristine sample. The images for irradiated samples show that as Se value increases, the grain size slightly decreases. The surface roughness in terms of root mean square (rms) was measured to be approximately 5 nm for pristine sample. Significantly, Ag ions at moderate fluence have reduced the surface roughness from 5 to 3.6 nm, while the corresponding reduction in rms roughness for Ni ions is much less (from 5 to 4.2 nm).

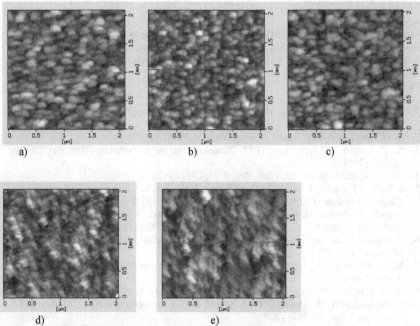

a)   b)   c)

d)   e)

Fig.4 AFM images of a) pristine, irradiated b) 80MeV Ni at 1×10^{12}, c) 100 MeV Ag at 1×10^{12}, b) 80MeV Ni at 3×10^{12} d) 100 MeV Ag at 3×10^{12} with the scan area of 2×2 μm^2. At higher fluence, a significant change of granular surface can be seen.

CONCLUSIONS

MOCVD grown Al $_{(1-x)}$In$_x$N/GaN HS were irradiated with 80 MeV Ni and 100 MeV Ag ions at varied fluence. HRXRD, RBS and AFM characterization techniques have been used for understanding the structural and surface morphology of pristine and irradiated samples. Philip's epitaxy simulation has been used for Indium composition of HS and is found to be 12 %. Random and simulated spectra of Al$_{1-x}$In$_x$N/GaN HS from RBS experiments are compared. It is evident from the RBS spectra that In and Al signals are completely separated from that of Ga signal. Increase in strain with increase of Se value has been noticed from HRXRD measurements. This means that there is an increase in Bragg angle of AlInN layer which could be attributed to ion beam induced lattice compression. Eventually, this work demonstrates that ion irradiation can create point defects which results in lattice expansion or compression. AFM images clearly show change of surface morphology with irradiation leading to reduction in granular sizes. Significantly, Ag ions at moderate ion fluence have resulted in lower surface roughness from 5 to 3.6 nm, while the reduction in rms roughness is much less for Ni ions irradiated samples (from 5 to 4.2 nm).

ACKNOWLEDGMENTS

G D and V S would like to thank CSIR for SRF. N S R would like to thank UGC-DAE-CSR for fellowship. G D would also thank DST, India for International travel grant support to attend the MRS meeting. We also thank Prof Arnab Bhattacharya of TIFR, India for helpful discussions.

REFERENCES

1. S D Lester, F A Ponce, M G Craford and , A Streigerwald, *Appl. Phys. Lett.* 66,1249(1995)
2. V. G. Debuk and A. V. Vozny, *Semiconductors.* **39**, 623(2005).
3. S. Y. Karpov, N. Podolskaya, I. A. Zhmakin, and A. I. Zhmakin, *Phys. Rev. B.* 70, 235203 (2004).
4. M. Ferhat and F. Bechstedt, *Phys. Rev. B.* 65, 075213 (2002).
5. R Butte, I F Carlin, E Feltin, M Gonschorek, S Nicolay, G Christmann, D Simeonov, A Catiglia, J Dorsaz, H J Buehlmann, , S. Christopoulos, G.B.H. von Hogersthal, A.J.D. Grundy, M. Mosca, C. Pinquier, M.A. Py, F. Demangeot, J. Frandon, P.G. Lagoudakis, J.J. Baumberg and N.J. Grandjean, *J. Appl. Phys. D.* 40, 6328 (2007)
6. A Dadgar, F Schulze, J Blasing, A Diez, A krost, M Neuburger, E Kohn, I Daumiller, and M Kunze, *Appl. Phys. Lett.* 85, 5400(2004)
7. W. H. Weber and R. Merlin, Editors, *Raman Scattering in Materials Science*, Springer, Berlin (2000).
8. B. D. White, M. Bataiev, L. J. Brillson, B. K. Choi, D. M. Fleetwood, R. D. Schrimpf, S. T. Pantelides, R. W. Dettmer, W. J. Schaff, J. G. Champlain and A. K. Mishra, *IEEE Trans. Nucl. Sci.*, 49, 2695 (2002)
9. J. F. Ziegler, J. P. Biersack, and M. D. Ziegler, The Stopping and Range of Ions in Matter, 2008, www.SRIM.org; originally based on TRIM code by J. F. Ziegler, J. P. Biersack, and U. Littmark, *The Stopping and Range of Ions in Solids* (Pergamon, New York, 1985).
10. M. A. G. Halliwell and M. H. Lyons, M. J. Hill, *J. Cryst. Growth.* 68, 523 (1984)
11. M. Mayer, Report IPP 9/113, Max–Planck-Institut für Plasmaphysik, Garching, Germany, 1997.

Mater. Res. Soc. Symp. Proc. Vol. 1354 © 2011 Materials Research Society
DOI: 10.1557/opl.2011.1344

5MeV Si Ion Modification on Thermoelectric SiO2/SiO2+Cu Multilayer Films

C. Smith[1] S. Budak[1], T. Jordan[2], J.Chacha[2], B. Chhay[3],
K. Heidary[2], R. B. Johnson [3], C. Muntele[1], D.ILA[4]

[1]Center for Irradiation of Materials, Alabama A&M University, Normal, AL USA
[2]Department of Electrical Engineering, Alabama A&M University, Normal, AL USA
[3]Department of Physics, Alabama A&M University, Normal, AL USA
[4]Department of Physics, Fayetteville St. University, Fayetteville, NC USA

Abstract

We prepared samples by electron beam physical vapor deposition EB-PVD followed
by ion bombardment. The samples were than characterized by photoluminescence
(PL), x-ray photoelectron spectroscopy (XPS). PL was used to characterize the
available energy states. XPS was used to determine the binding energies. The ML's
are comprised of 100 alternating layers of SiO_2/SiO_2+Cu.

*Corresponding author: C. Smith; Tel.: 256-372-5866; Fax: 256-372-5855;

Email: cydale.smith@cim.aamu.edu

1. INTRODUCTION

In previous studies we have demonstrated the improved thermoelectric properties of SiO_2
MLs embedded with various nanoparticles, such as Au, Ag and Cu [9-12].
ML structures are fabricated to reduce the phonon conduction through the sample, thus
increasing the figure of merit. The efficiency of the thermoelectric devices and materials is
determined by the figure of merit ZT [9]. The figure of merit is defined by $ZT = S^2\sigma T / \kappa$,
where S is the Seebeck coefficient, σ is the electrical conductivity, T is the absolute
temperature, and κ is the thermal conductivity [10, 11]. Increasing S can increase ZT, by
increasing σ, or by decreasing κ. In this study we have reported on the growth of
SiO_2/SiO_2+Cu multi-layer

We define ML as polycrystalline layers on the order of 10 nanometers that exhibit non bulk
properties. Koeler described a superlattice structure as a pair of layers of materials that has
single crystalline structure of 10nm thickness and less. Typically, the ML structures are
fabricated between 5 to 10 nanometers in order to take advantage of the quantum confinement
properties.

Quantum dots (QDs) have been used to increase the performance of devices for
numerous applications. Reed first coin the term quantum dot to describe semiconductor
particle that had a

Ion implantation is used to modify the ML region without damage. The stopping
powers of the ion traveling through the layers can be tailored to maximum the electronic
stopping power within the layer while minimizing the nuclear stopping power.

Cluster Formation

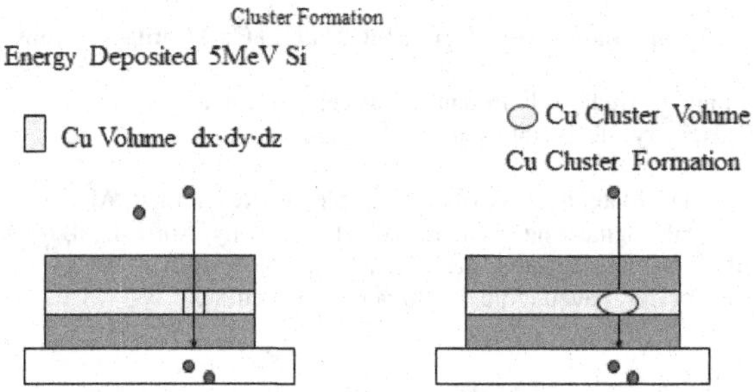

Figure 1. Multilayer thin films

2. EXPERIMENTAL

We have deposited the 100 alternating layers of SiO_2/SiO_2+Cu nana-layers films on silicon substrates at the total thickness of 382 nm. Figure 1. shows the fabricated ML films that were fabricated by using two Telemark electron beam 4-pocket guns. After EB-PVD we bombarded the samples with 5 MeV Si ions. The ion bombardment was performed with the 5SDH National Electrostatic Corp. Pelletron ion beam accelerator at the Howard J. Foster Center for Irradiation of Materials at Alabama A&M University. The fluences used for the bombardment were $1x10^{12} ions/cm^2$, $5x10^{12} ions/cm^2$, $1x10^{13} ions/cm^2$, $5x10^{13} ions/cm^2$, and $1x10^{14} ions/cm^2$. The RBS measurements were also done on the same 5SDH as mentioned previously. 2.1 MeV He was used as energy and projectile respectfully.

The photoluminescence and XPS were measured on a Varian Cary Eclipse spectrophotometer and Mantis Instruments respectfully.

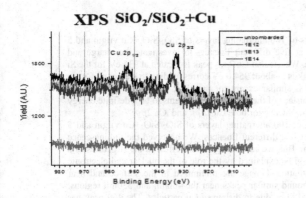

Figure 2. X-ray Photoelectron Spectroscopy

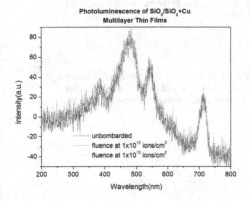

Figure 3. Photoluminescence of Multylayers

3. RESULTS AND DISCUSSION

Fig. 2 shows the XPS measurements of 100 alternating layers of SiO_2/SiO_2+Cu virgin and 5 MeV Si ion bombarded thin films at five different fluences to investigate the charge and electronic states of the implanted Cu. We see that Cu $2p_{3/2}$ peak is located at 934 eV for the Si bombarded Cu samples. Cu $2p_{1/2}$ peak is at about 954.0 eV corresponding to elemental Cu^0 or Cu^{1+}. The intensity for all fluences is similar except for the 1E12 ions/cm^2. This could be caused by excessive damage at the surface of the sample. There seems to be a definite energy shift in the spectrum as function of ion fluence for both Cu $2p_{1/2}$ and Cu $2p_{3/2}$.

Fig. 3 shows the PL measurements of 100 alternating layers of SiO_2/SiO_2+Cu virgin and 5 MeV Si ion bombarded thin films at five different fluences. We observed the luminescence bands at 460 nm, 550nm and 700 nm. But, we saw no significant effect in the structures as a function of the ion fluence. This could be explained by the role of the nuclear and electronic stopping powers. The effects of radiation damage are minimized in the surface and near surface regions. Other groups have found similar peaks near the 460nm and 550nm regions. The 550nm peak has been mentioned to be due to disbursed Cu particles. The 460 peak has been attributed to the electronic transitions, from the conduction band to Oxygen vacancy[15]. The peak at 700 can be possibly be related to nonbridging oxygen hole centers. The results are in good agreement with work of [15].

4. CONCLUSION

We have investigated the surface and photoluminescence properties of 100 alternating layers of SiO_2/SiO_2+Cu virgin and 5 MeV Si ion bombarded thin films at five different fluences. We saw no significant changes as function of ion fluence. This is due to the minimum impact of 5 MeV ion beam bombardment at the surface of the films.

Acknowledgement

Research sponsored by the Center for Irradiation of Materials (CIM), National Science Foundation under NSF-EPSCOR R-II-3 Grant No. EPS-0814103, DOD under Nanotechnology Infrastructure Development for Education and Research through the Army Research Office # W911 NF-08-1-0425.

References

1. L. L. Chang, L. Esaki, and R. TSu, Appl. Phys. Lett. 24,593 (1974)
2. Mehmet Arik, Jim Bray, and Stanton Weaver, Proc. of SPIE Vol. 7679 (2010) 76791F-1-18.
3. Zhen Xiong , Xihong Chen, Xueying Zhao, Shengqiang Bai, Xiangyang Huang, Lidong Chen, Solid State Sciences 11 (2009) 1612–1616.
4. Tao Li, Guangfa Tang, Gauangcai Gong, Guangqiang Zhang, Nianping Li, Lin Zhang, Applied Thermal Engineering 29 (2009) 2016-2021.
5. T. C. Harman, P.J. Taylor, M.P. Walsh, B.E. LaForge, Science 297(2002) 2229-2232.
6. J.L. Liu, A. Khitun, K.L. Wang, T. Borca-Tasiuc, W.L. Liu, G. Chen, D.P. Yu, Journal of Crystal Growth 227-228 (2001) 1111-1115.
7. G. Slack, in CRC Handbook of Thermoelectrics (Ed: D.M. Rowe), CRC Press (1995) pp.407-440.
8. T.M. Trit, Ed., Semiconductor and Semimetals 71 (2001)
9. Chin-Hsiang Cheng , Shu-Yu Huang , Tsung-Chieh Cheng, International Journal of Heat and Mass Transfer 53 (2010) 2001–2011.
10. B. Zheng, S. Budak, C. Muntele, Z. Xiao, C. Celaschi, I. Muntele, B. Chhay, R.L. Zimmerman, L.R. Holland, D. Ila, Materials in Extreme Environments, Materials Research Society, vol. 929, 2006, p. 81.
11. S. Budak, C. Muntele, B. Zheng, D. Ila, Nuc. Instr. and Meth. B 261 (2007) 1167.
12. S. Guner, S. Budak, R. A. Minamisawa, C. Muntele, D. Ila, Nuc. Instr. and Meth. B 266 (2008) 1261.
13. S. Budak, S. Guner, R. A. Minamisawa, and D. ILA, Surface and Coating Technology 203 (2009) 2479-2481.
14. J. Chacha, S. Budak, C. Smith, M. Pugh, K. Ogbara, K. Heidary, R. B. Johnson, C. Muntele, D.ILA, Mater. Res. Soc. Symp. Proc. Vol. 1267 © 2010 Materials Research Society 1267-DD05-15.
15. N. Umeda, N. Kishimoto, Y. Takeda, C.G. Lee, V.T. Gritsyna, Nucl. Instrum. Methods B 166–167 (2000) 864.

Mater. Res. Soc. Symp. Proc. Vol. 1354 © 2011 Materials Research Society
DOI: 10.1557/opl.2011.1458

Study of scalable IBS nanopatterning mechanisms for III-V semiconductors using in-situ surface characterization

Jean Paul Allain[1,2], Osman El-Atwani[1,2], Alex Cimaroli[1], Daniel L. Rokusek[1], Sami Ortoleva[1], Anastassiya Suslova[1],

[1]Purdue University, West Lafayette, IN 47907, USA
[2]Birck Nanotechnology Center, West Lafayette, IN 47907, USA

ABSTRACT

Ion-beam sputtering (IBS) has been studied as a means for scalable, mask-less nanopatterning of surfaces. Patterning at the nanoscale has been achieved for numerous types of materials including: semiconductors, metals and insulators. Although much work has been focused on tailoring nanopatterning by systematic ion-beam parameter manipulation, limited work has addressed elucidating on the underlying mechanisms for self-organization of multi-component surfaces. In particular there has been little attention to correlate the surface chemistry variation during ion irradiation with the evolution of surface morphology and nanoscale self-organization. Moreover the role of surface impurities on patterning is not well known and characterization during the time-scale of modification remains challenging. This work summarizes an *in-situ* approach to characterize the evolution of surface chemistry during irradiation and its correlation to surface nanopatterning for a variety of multi-components surfaces. The work highlights the importance and role of surface impurities in nanopatterning of a surface during low-energy ion irradiation. In particular, it shows the importance of irradiation-driven mechanisms in GaSb(100) nanopatterning by low-energy ions and how the study of these systems can be impacted by oxide formation.

INTRODUCTION

It is well known that many shapes and sizes of nanostructures can be formed via ion beam sputtering (IBS) techniques. Numerous techniques are known that pattern surfaces including: block co-polymer self-assembly. Ion irradiation can be operated at very low energies (e.g. below the threshold energy for displacement damage) and can introduce new processing pathways not offered by traditional thermodynamic self-assembly approaches. IBS also provides the flexibility to tailor both the surface concentration and surface nanopatterning by changing the ion fluence, incident ion angle, ion energy and co-implantation of ion beams. With device features approaching characteristic lengths of the order of several monolayers (~ 1-2 nm), irradiation at low energies near the threshold regime becomes invaluable for ion-irradiation based nanopatterning. There are two primary reasons for this requirement. One, plasma-based processing of materials continues to be a reliable, efficient and versatile method to modify materials. Although very high-energy ion beams can modify the surface of materials via electronic energy losses, ion-beam accelerators cannot be cheaply integrated into existing materials processing tools. Second, the stopping power is such that to modify only the top few nm of a surface, extremely low-energy ions must be used. Thus the work presented here focuses on energies that range between 50 and 1000 eV.

The role of metal impurities on pattern formation is not well understood [1-6]. Moreover, the role of oxygen in the formation mechanism of the nanostructures is almost neglected. The oxide layer on top of gallium antimonide substrates plays an important role in the differential sputtering mechanism. For example, oxides are known to erode much more slowly than their metallic counterparts. Gallium oxide is very strongly bonded compared to the bonding of antimony oxide, and thus, differential sputtering of antimony will be enhanced. The formation of gallium oxide is improved due to the following reaction:

$$Sb_2O_3 + 2GaSb \rightarrow Ga_2O_3 + 4Sb \qquad \text{Equation 1}$$

In this study, two different sets of experiments are performed on gallium antimonide substrates. In the first set of experiments, gallium antimonide substrates are irradiated at the low and high energies with argon ion beams. In-situ X-ray photoelectron spectroscopy (XPS) and low energy ion scattering spectroscopy (ISS) characterization are performed on the samples at different fluencies. The results are supplemented with ex-situ Scanning Electron Microscopy (SEM) and Trasnmission Electron Microscopy (TEM) morphology characterization of the irradiated samples. In another set of experiments, however, similar work was performed on gallium antimonide but under ex-situ conditions, where the samples were exposed to atmosphere before XPS and ISS characterizations. All irradiations and chemical characterizations were performed in the Particle and Irradiation Interaction of Hard and Soft Matter (PRIHSM) facility in the School of Nuclear Engineering at Purdue University. XPS results indicated the abundance of gallium oxide over antimony oxide. The antimony oxide peak vanished much before the gallium oxide peak.

Quantification of the results showed large differences between the results of the ex-situ and in-situ experiments. The results demonstrate that it is crucial to perform the nanostructures formation mechanism studies under in-situ conditions. Results from experiments performed with exposure to air (e.g. ex-situ) are in most cases misleading. As an example, irradiation of gallium antimonide using an argon ion beam at 100eV was performed. In-situ characterization using ISS and XPS showed 45% and 35% of gallium relative concentration respectively. When the samples are exposed to atmosphere and then characterized with ISS and the XPS, the results changed to 54% (ISS) and 45% (XPS) of gallium relative concentration. The difference between the XPS and ISS in both cases is related to the difference in probing depth of both techniques. During the initial irradiation stages, when an amorphous layer is being formed, gallium preferentially captures oxygen due to the native oxide layer that exists. Ex-situ results showed more gallium oxide on the surface. This is due to preferential reduction of oxygen on the surface. The results can show more gallium on the surface if exposed more to atmosphere for a longer time due to the reaction in equation 1.

EXPERIMENTAL SETUP

III-V compound semiconductor experiments were performed using the PRIHSM (Particle and Radiation Interactions with Hard and Soft Matter) facility. The facility enables characterization of multi-component surfaces at several depth scales (0.1-2.0 nm) under irradiation by well-controlled low-energy ions. PRIHSM provides two primary surface characterization techniques used in this study. One is low-energy ion scattering spectroscopy (LEISS) and the second is x-ray photoelectron spectroscopy (XPS). Both are conducted *in-situ* allowing direct correlation with the irradiation conditions of the ion beam. Furthermore the tilt angle between the diagnostic ion beam and the detector can be modified to suit the ion-induced nanostructures. This allows for probing of surface nanopatterned high-aspect ratio structures that may have local variations of surface chemistry that correlated with their morphology. Fig. 1 illustrates the setup of the *in-situ* facility known as PRIHSM (Particle and Radiation Interaction with Hard and Soft Matter) that was designed by Allain and Rokusek. A similar *in-situ* system was designed by Allain et al. known as IMPACT and its details are included in an earlier paper [7].

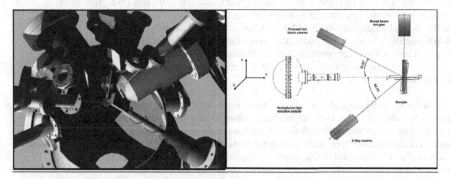

Figure 1. An illustration of the PRIHSM experimental configuration. On the left panel is a 3D rendering of the PRIHSM sample setup. The chamber was custom-designed for *in-situ* surface characterization experiments allowing for multiple, complementary techniques. The right panel shows the angles corresponding to the locations of the analyzer, ion gun location and sample normal.

Irradiations are conducted at energies of 100, 500 and 1000 eV primarily using Ar$^+$ ion beams, although ion some cases He and Ne ion beams have been used. The ions are incident normal to the sample surface as depicted in the right frame of Fig. 1. Active cooling maintains the target at a temperature near 25-30 C. A VG Scienta 2000 2D hemipherical sector energy analyzer is used to obtain energy spectra of X-ray induced photoelectrons or low-energy ions scattered from the sample surface. For experiments designated "*ex-situ*" the surface concentration is measured after each dose exposure. For those "*in-situ*" experiments, the surface concentration is measured throughout the ion fluence without any air exposure between measurements.

RESULTS AND DISCUSSION

Ex-situ III-V compound semiconductor nanopatterning

The surface concentration of GaSb(100) samples irradiated with Ar$^+$ ions at normal incidence and energies between 50-1000 eV for fluences up to several times 10^{18} cm^{-2} are presented in this section. The first set of experiments is critical in that the characterization of GaSb samples is conducted *after exposure to ambient air*. As will be shown, the role of oxide on nanopatterning and surface chemistry dictates the resulting patterning behavior. The second set of experiments consists of *in-situ surface characterization* without any exposure to ambient air.

Total fluences of $2-5\times10^{17}$ cm^{-2} Ar$^+$ were delivered to GaSb samples. Prior to Ar$^+$ bombardment, each sample was cleaned for 15 minutes with a 1 keV Ar$^+$ beam at 30-degrees from normal. A series of in-situ XPS scans were performed before, during, and after Ar$^+$ cleaning and bombardment of the samples. XPS spectra provide information pertaining to the elemental state of the sample in the near-surface region (0.5-5.0 nm), which can be used to examine Ar$^+$-induced damage and modification mechanisms. Figure 2 shows the Ga/Sb surface ratio for Ar$^+$ incident on GaSb (100) at 1 and 1.5 keV and at 0-, 30-, and 75-degrees from normal. Also plotted are simulation results obtained from the DYNAMIX code [8] for 1 keV Ar$^+$ at 0-, 30-, and 75-degrees. DYNAMIX (DYNamic transport of multi-Atom material MIXing) is a dynamic BCA Monte Carlo code. Data were taken from snapshots for increasing fluence of concentration versus depth profiles until equilibrium was reached. The depth of penetration of the implanted ions was calculated to be 2.0, 1.5, and 1.0 nm for 0-, 30-, and 75-degrees, respectively.

At moderate fluences above 10^{16} cm^{-2} preferential Sb sputtering which leads to a slight Ga surface enrichment seems to be the mechanism that governs the surface behavior. Overall, the Ga/Sb surface ratio increases with dose. Intensities of the Ga2p and Sb3d peaks from in-situ XPS spectra at each fluence level were used to calculate the Ga/Sb ratio. However, this was also done for each fluence data point the sample was exposed to air. As the next section will show exposure to air in fact compromised the surface chemistry and led to an incorrect initial conclusion on the governing mechanism.

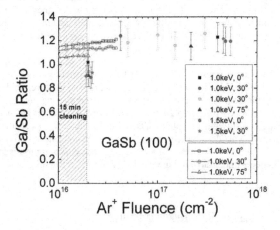

Figure 2. Ga/Sb ratio vs Ar$^+$ fluence for various incident angles and energies.

AFM analysis was conducted for Ar bombarded GaSb samples. Roughness measurements of a virgin GaSb sample showed area RMS roughness of 2.0 nm and average height of 3.1 nm. Figure 3 shows nanostructures in the form of islands. These islands were produced by 1500 eV Ar$^+$ at normal incidence. An average island height of 2.64 nm and standard deviation of 0.56 nm over 40 sampled islands were measured, with area RMS roughness of 0.9796 nm. At 75-degrees with respect to surface normal, 1000 eV Ar$^+$ ions created periodic ripple structures rather than the islands.

In-situ III-V compound semiconductor nanopatterning

The data presented in the previous section consisted of surface characterization conducted *ex-situ* for each fluence exposure with Ar$^+$ ions. To elucidate on the behavior of Ga and Sb atoms during irradiation a novel *in-situ* surface characterization tool is used. The PRIHSM facility allows the *in-situ* measurement of GaSb(100) surfaces irradiated with Ar$^+$ ions sequentially without breaking vacuum. As Fig. 3 shows (compared to data in Fig. 2) there is a signficant fluence-dependent behavior of the Ga concentration. The Ga relative concentration is plotted vs fluence from both the LEISS and XPS data. The LEISS data probes predominantly the ultrathin 2-nm amorphous phase and XPS both the amorphous phase and crystalline phase of GaSb.

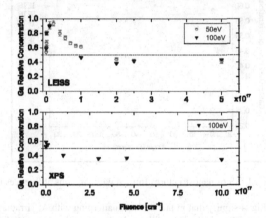

Figure 3. The LEISS and XPS data adapted from El-Atwani et al [8] as a function of Ar+ irradiation fluence.

The data in Fig. 3 clearly show that after 10^{17} cm^{-2} fluence the oxide is removed and ion-induced nanopatterning ensues with segregation of Sb atoms to the surface. This is in contrast to the predominant Ga surface concentration evidenced in the *ex-situ* data of Fig. 2.

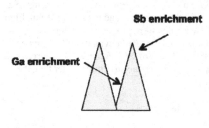

Sb enrichment

Ga enrichment

Figure 4a. Schematic of AR-LEISS data.

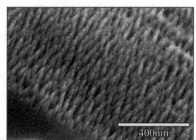

Figure 4b. SEM of GaSb nanostructures formed by 1-keV Ar$^+$ irradiation.

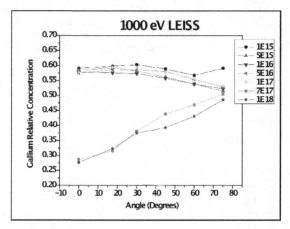

Figure 4c. 1-keV AR-LEISS of nanostructures formed with Ar$^+$ ions at fluences up to 10^{18} cm^{-2}

The results shown in Fig. 4 signify that in fact the nanopatterning with Ar$^+$ irradiation leads to a net amount of Sb enrichment on the surface of the nanostructures that appear to form near 5 x 10^{17} cm^{-2} fluence. The correlation between the nanopatterns and surface concentration between Ga and Sb atoms during irradiation was also studied by conducting angle-resolved LEISS measurements capable of probing the sides of the nanostructures formed by Ar$^+$ irradiation.

Fig. 4 shows the angle-resolved LEISS (AR-LEISS) data for the case of 1-keV Ar$^+$ irradiation. Each angle-dependent Ga relative concentration curve is measured *in-situ* without exposure to air. The lines in Fig 4c are shown to "guide the eye" and the error in the data is about 5% or the size of the symbol. Note that after 7 x 10^{17} cm^{-2} a dramatic change in the angle-resolved LEISS data takes place. The very small angles, which correspond to predominant measurement of the tip of the conical nanostructure, suggest that it is heavily Sb-enriched. Also note that the angle-dependent data do not follow a straight horizontal line (which indicates a constant concentration from the tip of the nanostructures to the sides or troughs of the conical shape). In fact, at high fluence this dramatic amount of Sb at the surface is indicative of an effective mass transport and redistribution mechanism.

CONCLUSIONS

Low-energy directed irradiation synthesis is used to modify GaSb surfaces in regimes far from equilibrium where surface roughening and surface smoothening mechanisms (sputtering and surface diffusion) are present. In particular, we correlate surface chemistry by *in-situ* measurements with nanopatterns diagnosed using SEM. In GaSb it is Sb atoms that dominate the surface. When exposed to air, Ga atoms reduce oxygen to form a layer on the surface. Moreover the AR-LEISS data imply that another mechanism may be responsible for patterning, given the apparent mass transport from the top of the conical nanostructures. PRIHSM is a versatile new experimental facility designed to study irradiation-induced changes in low-dimensional state materials.

ACKNOWLEDGEMENTS

We would like to thank Brandon Holybee for AFM results. This work is supported in part by the Department of Energy's 2010 Early Career Award DE-SC0004032 and the NRC (NRC-38-08-948) Faculty Development grant. We also would like to acknowledge helpful discussions with Stefan Facsko, Michael Aziz and Mark Bradley.

REFERENCES

1. J. Erlebacher and M. J. Aziz, *Phys. Rev. Lett.* **82**, 2330 (1999).
2. S. Fascko, T. Dekorsy, C. Koerdt, C. Trappe, H. Kurz, A, Vogt, H. L. Hartnagel, *Science* **285**, 1551 (1998).
3. T. Bobek, S. Facsko, H. Kurz, T. Dekorsy, M. Xu, and C. Teichert, *Phys. Rev. B* **68**, 085324 (2003).
4. B. Ziberi, F. Frost, B. Rauschenbach, and Th. Hoche, *Appl. Phys. Lett.* **87**, 033113 (2005).
5. F. Frost, B. Ziberi, A Schindler, and B. Rauschenbach, *Appl. Phys. A* **91**, 551 (2008).
6. M. Cornejo, J. Völlner, B. Ziberi, F. Frost and B. Rauschenbach, Fabrication and Characterization in the Micro-Nano Range, in Advanced Structured Materials, 10, Ed. F.A. Lasagni, A.F. Lasagni, Springer-Verlag Berlin, Heidelberg 2011. pp 69-94
7. J. P. Allain, M. Nieto, M. R. Hendricks, P. Plotkin, S. S. Harilal, and A. Hassanein, *Rev. Sci. Instrum.* **78**, 113105 (2007).
8. G. Hou, MS Thesis 2010, Purdue University, advisor: J.P. Allain.
9. O. El-Atwani, J.P. Allain, S. Ortoleva, *Nucl. Instrum. Method. B*, In Press 2011, doi:10.1016/j.nimb.2011.01.067.

Mater. Res. Soc. Symp. Proc. Vol. 1354 © 2011 Materials Research Society
DOI: 10.1557/opl.2011.1209

Modeling of approximated electron transport across ion beam patterned quantum dot nanostructures

J. Lassiter[1], J. Chacha[2], C. Muntele[3], S. Budak[2], A. Elsamadicy[1], D. ILA[3, 4]

[1]Department of Physics, University of Alabama Huntsville, Huntsville, AL USA
[2]Department of Electrical Engineering, Alabama A&M University, Normal, AL USA
[3]Center for Irradiation of Materials, Alabama A&M University, Normal, AL USA
[4]Department of Physics, Alabama A&M University, Normal, AL USA

Abstract

High energy ion beams are used to modify co-deposited nanolayer films of alternated materials (e.g. insulator and metal, two different semiconductors, even more complex arrangements) to form nanodots through localized nucleation. The particular application being considered here is for high efficiency thermoelectric conversion systems. The performance of a thermoelectric converter is generally given by the figure of merit, ZT, which is a function of the Seebeck coefficient, electrical conductivity, and thermal conductivity. A high performance device would have a maximized electrical conductivity and a minimized thermal conductivity (maximum electron transport, minimal phonon transport). The current models of electron and phonon transportation through 1D, 2D, 3D quantum regimented structures assume an infinitely repetitive perfect structural "cell", with complicated algorithms requiring intensive computing power. The main focus for this modeling effort is to reduce the three-dimensional problem to a single dimensional approximation without sacrificing the quality of the result. The nanostructure being investigated has Si and Ge quantum dots arranged with perfect periodicity in all three Cartesian directions, with the heat and electricity flow monitored in the z-direction (cross-plane as initially layered). Non-Equilibrium Green's Functions Formalism (NEGF) is the mode for calculating the theoretical electrical properties assuming a one-dimensional quantum well arrangement in the z-direction (with finite boundaries in the x and y directions) with tight binding (nearest neighbor approximation). The results are to be compared with experimental measurements on such structures.

Introduction

The dimensionless figure of merit, ZT, is conventionally utilized to quantify the efficiency of a thermoelectric device. Its dependence on the fundamental parameters of the materials is shown in equation (1), with S, σ, T, k, Z being the Seebeck coefficient, electrical conductivity, average temperature between the two contacts, thermal conductivity, and figure of merit, respectively.

$$ZT = \frac{S^2 \sigma T}{k} \tag{1}$$

The motivation of this work was to understand and model theoretically the mechanics of electronic conduction effects on nanostructured thermoelectric systems and in the future couple phonon conduction and mechanical properties of our stacked thin films to make a more realistic model. This model will be utilized to choose materials that are more promising for further experimental development and prototyping of a final high ZT system.

Model

The modeling effort was inspired by the works of A. Bulusu and D. G. Walker [1] except that the dielectric constant of the system was calculated using the formulation [2, 3]:

$$\varepsilon_{in} = 1 + \frac{\left(\varepsilon_{bulk} - 1\right)}{\left[\left(1 + 1.84/d\right)^{1.18}\right]} \tag{2}$$

Instead of using the bulk dielectric constant, the layer will have the permittivity of spherical quantum dots, which is a function of the bulk dielectric constant and the diameter of quantum dot [3]. Delerue, Lannoo and Allan contend that the change in dielectric constant only occurs on the surface due to breaking of polarizable bonds in that region [2], but in a later paper Delerue and Allan came to the conclusion that nanostructuring of thin films influences the dielectric constant because of local field effects, and size of quantum dots [3]. The model relied on a system of thin film layers assumed to have the same effective dielectric correction, hence our use of the equation above. The physical system being modeled has a 3 layer Si / Ge / Si system with two source and drain contacts (Figure 1a, b) with layer thicknesses of 1 nm, 2 nm, and 1 nm respectively discretized in the z-direction.

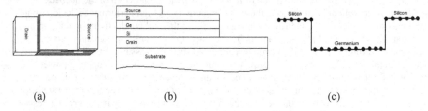

(a)   (b)   (c)

Figure 1. 3D (a) and 2D (b) representations of the structure considered, and the discretization assumed in the modeling (c).

The discretization of the layer (Figure 1c) followed the convention used in [1] with an idealized breakdown showing how the original cross-sectional device will look and how the layers will be discretized (regardless of the actual atomic structure and orientation). The source to drain voltage range was from 0 to 0.1V with increments of 0.5 mV, for a total of 201 voltage points. The temperature was biased from drain to source, with cold representing the baseline and hot side incremented from the baseline to create the temperature gradient shown in Table 1.

Table 1. Drain to Source temperature profile and test situations.

Baseline T (K)	ΔT (K)			
300	0	10	20	30
800	0	10	20	30
1300	0	10	20	30

The Non-Equilibrium Green's Function (NEGF) formalism has been developed to treat transport effects in nanoscale devices, since the Boltzmann Transport Equations (BTE) works best in the classical and semi-classical regimes. NEGF is a versatile formalism that is capable of modeling ballistic transport, reflection, and even particle interactions [4].

NEGF is derived from the Schrodinger equation in the Helmholtz form (Helmholtz equation) where the delta function (impulse response function) is taken as a source term. Physically, the impulse response function pulses the electrical charge, thus describing a flow of discrete carriers rather than a continuous flow of charge. The Green's Function is defined below in terms of the delta function (or identity matrix) below:

$$(\nabla^2 + k^2)G(r) = \delta(r - r')$$ (3)

where $(\nabla^2 + k^2)$ is a linear operator. Equation 3 is typically written as:

$$(E - H)G(z,z') = \delta(z - z')$$ (4)

where $\delta(z - z')$ is the impulse response function.

When applied to $H\psi_a = E\psi_a$, this yields $H_{Effective}\psi_a = E_a\psi_a$ [1, 5]. The periodicity from invoking the Kronig-Penney model emerges in $E(k) = E_c + \frac{\hbar^2}{2m_z}\left(\frac{n\pi}{L_z}\right)^2$ [1]. Here k is given to be the wave vector, which gives the direction of propagation. Current travels in the z-direction where confinement is assumed. Tight binding is also assumed, which means only the diagonal elements of the matrix are taken into account [1, 4]. From the Green's functions, the spectral function $A = i(G - G^\dagger)$ and Landauer's formula were utilized to determine the density of states and current, respectively.

Results and Discussions

The code outputs current densities for each subband as functions of the voltage steps. Electrical conductivity is thus calculated as voltage and temperature dependent:

$$\sigma = \frac{I_{device} \cdot J_0}{V_{bias}}$$ (5)

Figure 2 shows the electrical conductivities calculated with Equation 5 for the tabulated (Table 1) test conditions. Notice that the electrical conductivity is voltage dependent, not only temperature dependent.

Seebeck coefficients are given by:

$$S = \frac{V_{Seebeck}}{\Delta T}$$ (6)

Seebeck coefficients are found to be inversely proportional to baseline temperature (Figure 3). The current base code [1] needs a bias across the system (source to drain) to calculate charge carriers flow (current density per subband). To find the Seebeck voltage, a temperature gradient needs to be applied such that it generates a charge carrier flow opposite to that generated by the applied bias. Current densities for voltage steps between preset limits are generated. The Seebeck voltage for a given ΔT is the voltage for which the net charge carrier flow cancels out.

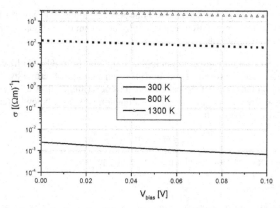

Figure 2. Calculated electrical conductivity as function of baseline temperature and applied voltage.

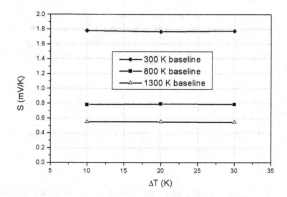

Figure 3. Seebeck coefficient values as function of temperature gradients and baseline temperatures.

Figure 4 shows the Seebeck voltage generated as a function of temperature. As expected, there is no Seebeck voltage generated for $\Delta T = 0$ K. Notice that the Seebeck voltage decreases with increasing of the baseline temperature for the same ΔT, most likely due to the higher state of thermal agitation of carriers, correlating with $\Delta T/T$.

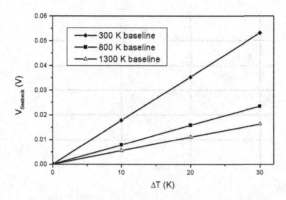

Figure 4. Seebeck voltage generated as function of temperature gradients and baseline temperatures.

Conclusions and future plans

Seebeck coefficient and electrical conductivity results are comparable with existing literature for both simulated and experimental conditions (i.e. confidence in the theoretical model and proper coding). The electrical conductivity is voltage dependent, not only temperature dependent.
Use of Matlab becomes time-consuming when discretization is increased, therefore we see the need to move to C++. We also need to modify the code to require temperature gradients only, without the need of bias voltage. Also, additional data processing will be added to the end side of the code, rather than just raw current densities. As a final improvement necessary, we need to include a repetition of the unit cell / device mentioned here, to account for the real experimental situation.

References

[1] Bulusu and D.G. Walker, "Quantum Modeling of Thermoelectric Properties of Si/Ge/Si Superlattices," IEEE Transactions on Electron Devices, vol.55, no.1, pp.423-429, Jan. 2008 doi: 10.1109/TED.2007.910574
[2] Delerue, M. Lannoo, and G. Allan "Concept of dielectric constant for nanosized systems", Phys. Rev. B 68, 115411, 2003
[3] Delerue and G. Allan, "Effective dielectric constant of nanostructured Si layers" Appl. Phys. Lett. 88, 173117 (2006), DOI:10.1063/1.2198814
[4] M. P. Anantram, M. S. Lundstrom, and D. E. Nikonov, "Modeling of Nanoscale Devices", Proceedings of the IEEE, v. 96, no. 9, pp. 1511 - 1550 (2008).
[5] M. Paulson "Non Equilibrium Green's Functions for Dummies: Introduction to the One Particle NEGF equations."arXiv:Cond-mat/0210519v2 [cond-mat.mess-Hall] 3 Jan 2006.

Mater. Res. Soc. Symp. Proc. Vol. 1354 © 2011 Materials Research Society
DOI: 10.1557/opl.2011.1082

RBS, XRD, Raman and AFM Studies of Microwave Synthesized Ge Nanocrystals

N Srinivasa Rao, A P Pathak*, G Devaraju, V Saikiran and S V S Nageswara Rao

School of Physics, University of Hyderabad, Central University (P.O), Hyderabad 500 046, India.

ABSTRACT

Ge nanocrystals embedded in silica matrix have been synthesized on Si substrate by co-sputtering of SiO_2 and Ge using RF magnetron sputtering technique. The as-deposited films were subjected to microwave annealing at 800 and 900^0C. Rutherford backscattering spectrometry (RBS) has been used to measure the Ge composition and film thickness. The structural characterization was performed by using X-ray diffraction (XRD) and Raman spectrometry. XRD measurements confirmed the formation of Ge nanocrystals. Raman scattering spectra showed a peak of Ge-Ge vibrational mode around 299 cm^{-1}, which was caused by quantum confinement of phonons in the Ge nanocrystals. Surface morphology of the samples was studied by atomic force microscopy (AFM). Variation of nanocrystal size with annealing temperature has been discussed. Advantages of microwave annealing are explained in detail.

Keywords: microwave heating, nanostructure, Ge

*Corresponding author E-mail: appsp@uohyd.ernet.in
Tel: +91-40-23010181/23134316, Fax: +91-40-23010181 / 23010227.

INTRODUCTION

Quantum confinement effects are known to play an essential role in optical absorption and emission processes in semiconductor nanocrystals since the energy band gap increases with decreasing particle size and electronic states become discrete with higher oscillator strength. In semiconductor nanocrystals the excitons are confined in all three spatial dimensions. As a result, they have different properties than those of bulk semiconductors. Besides, the surface electronic states also affect both electronic and optical properties of semiconductor nanocrystals due to large surface-to-volume ratios [1–3]. For these reasons, enormous amount of research effort has been made worldwide, to synthesize and study these materials. Particularly Si and Ge nanocrystals embedded in a dielectric matrix have been demonstrated as potential candidates for the fabrication of optoelectronic, photovoltaic and nonvolatile memory devices [4–6]. Different annealing techniques have been used for synthesis of Ge nanocrystals such as furnace annealing [7] and rapid thermal annealing [8]. In addition to above annealing methods, microwave annealing has been recognized to be a good technique for synthesizing nanocrystals, having its own advantages. **Here, we present a brief report on the principal features and conclusions from our recently published paper [9] "Growth and characterization of nc-Ge prepared by microwave annealing" which was presented as a poster in the Spring, 2011 MRS meeting.** Structure of the synthesized films has been evaluated by XRD and Raman spectroscopy. Surface morphology of the films was studied by AFM. RBS was used to estimate composition of Ge in as-deposited samples.

EXPERIMENTAL DETAILS

The composite thin films of Ge and SiO_2 were deposited by RF-magnetron sputtering technique on Si substrates. These as deposited films were subjected to microwave annealing at 800 and 900^0C. The microwaves used in these experiments were produced using a 1.3 kW, 2.45 GHz single mode applicator. Proper amount of silicon carbide was used as microwave susceptor around the sample within the insulating package to preheat the sample and compensate the heat loss from the sample. The composite films were analyzed by Rutherford backscattering spectrometry (RBS) using the HVE 1.7 MV Tandetron accelerator. For RBS, 2 MeV He^+ beam was used as a projectile and the scattered particles were detected by a surface barrier Si detector with scattering angle of 165° from the incident beam direction. Raman scattering spectra of the films were obtained before and after annealing in backscattering configuration using a 514.5 nm Ar^+ laser excitation source. X-ray profiles were obtained with CuK_α X-ray, $\lambda=0.154nm$ in a glancing angle incidence geometry. The atomic force microscopy (AFM) measurements were carried out using Model SPA400, Seiko Instruments Inc.

RESULTS AND DISCUSSION

Fig. 1 shows the XRD spectra of as deposited and annealed samples at 800 and 900^0C. As deposited sample shows amorphous nature whereas annealed ones show crystalline nature with three peaks of Ge (1 1 1), (2 2 0) and (3 1 1), which indicates the formation of nc-Ge (nanocrystalline germanium). It is also observed that with the increase of annealing temperature, the XRD peaks become sharper and the full width at half-maximum (FWHM) of each peak decreases. This means that the average particle size of nc-Ge increases. Size of the nanocrystals calculated using Scherrer formula increases from 11 nm to 23 nm when the annealing temperature increases from 800 to 900^0C.

Raman scattering is expected to be sensitive not only to quantum confinement, but also to disorder, strain and structural characteristics of the system. Fig. 2 shows Raman spectra of the as-deposited and the annealed samples. It can be seen that the as-deposited sample exhibits a broad band which is a characteristic feature of amorphous Ge. However, for the annealed sample, a sharp Raman peak was observed at around 299 cm^{-1}, which is different in comparison with that of the bulk Ge, indicating Ge crystallization [10]. For bulk crystalline Ge, the Raman spectrum exhibits Lorentzian line at 300 cm^{-1} with a width of 3 cm^{-1}. The origin of the shift and asymmetrical broadening of the Raman spectra are due to the confinement of phonons and are also characteristic of nanocrystals [11].

The feature located around 420 cm^{-1} in annealed sample at 900^0C is due to Ge-Si vibrations. This Ge-Si signal is due to alloying of Ge with Si at the surface as a result of annealing. We can also see that Ge crystallinity increases with the increase of annealing temperature. Generally, when the annealing temperature increases and reaches a threshold level then the partial structure of the as-deposited film transforms from amorphous to crystalline. In other words, the amorphous and crystalline phases coexist. But, when the annealing temperature

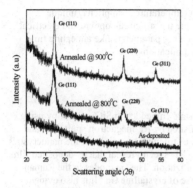

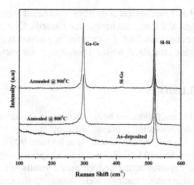

Figure 1. GIXRD spectra of as-deposited and annealed samples

Figure 2. Raman spectra of as-deposited and annealed samples

increases further, then more of the amorphous structure in the as-deposited film will be transformed into crystalline structure, as reflected by the intensity of the Raman peak. In as-deposited sample, there is no signature of nanoparticles, but annealing results in formation and growth of nanoparticles. Due to high temperature annealing, the diffusivity of Ge atoms and small nanocrystals increases such that the nucleation rate is enhanced leading to a higher density of nanocrystals. Since nc-Ge crystallization temperature is lower than Ge melting point temperature, Ge clusters and nanocrystals were formed by diffusion of Ge atoms or clusters inside SiO_2 matrix.

Various annealing methods have been used to prepare nanocrystals like furnace processing, Rapid Thermal Annealing (RTA), and Laser Annealing. But, each of these methods has its own limitations. It is well known that furnace annealing takes a few hours to complete the material processing. To avoid surface deterioration caused by slow heating and cooling rates of conventional resistive heating furnaces, new ultrafast annealing techniques need to be explored. Similarly, in Laser Annealing method, the spot size is very small. Hence, it requires a lot of time to cover entire surface area of wafer. In this process there is high probability for overlapping or missing some part of the wafer. Halogen lamp and laser-based rapid thermal processing techniques suffer from problems such as a possible limitation on the maximum achievable annealing temperature, surface melting, a large residual defect density, and redistribution of the implants. But microwave heating may solve many of these problems. Microwave heating is a process in which the materials couple with microwaves, absorb the electromagnetic energy volumetrically, and transform it into heat. In conventional annealing methods the heat is transferred between objects by the mechanisms of conduction, radiation and convection. In conventional heating, there is a temperature gradient from the surface to the inside since the material's surface is first heated followed by the heat moving inward. However, microwave heating generates heat within the material first and then heats the entire volume [12]. This heating mechanism is useful for enhanced diffusion processes, reduced energy consumption, very rapid heating rates and considerably reduced processing times, improved physical and mechanical properties, decreased sintering temperatures, simplicity, unique properties, and lower environmental hazards. One can tune the structural, optical and electrical properties of nanocrstals by varying the microwave annealing parameters like annealing time, temperature, and RF deposition parameters for various applications.

CONCLUSIONS

In conclusion, we have synthesized Ge nanocrystals embedded in SiO_2 matrix by using microwave annealing process. The films were first deposited by RF magnetron sputtering and subjected to microwave annealing at 800 and 900^0C. The structural properties of the nanocrystals were characterized by XRD and Raman spectroscopy. XRD spectra reveal that as-deposited film shows amorphous nature and annealing results in crystallization. The Ge modes in the Raman spectra of annealed samples were indicating the existence of crystalline Ge. It is also evident from XRD and Raman that the crystallinity increases with increasing annealing temperature. The estimated nanocrystal size is found to vary with annealing temperature. The variation of crystallite size as a function temperature and advantages of microwave annealing have been discussed.

ACKNOWLEDGEMENTS

A P P thanks Center for Nano Technology for DST-Nano project. N S R would like to thank UGC DAE-CSR for fellowship. G D and V S thank CSIR for SRF. We also thank Dr K C James Raju and Mr A Rambabu for use of their microwave facility.

REFERENCES

1. S. Takeoka, *Phys. Rev. B* **58,** 12 (1998)
2. L. Brus, *Appl. Phys. A* **53,** 465 (1991)
3. Y.X. Jie, Y.N. Xiong, A.T.S. Wee, C.H.A. Huan, W. Ji, *Appl. Phys. Lett.* **77**, 3926 (2000)
4. R J Walters, G I Bourianoff and H A Atwater, *Nat. Mater.* **4**, 143 (2005)
5. G Conibeer, M A Green, R Corkish, Y Cho, E C Cho, C W Jiang *et al.,Thin Solid Films* **511/512,** 654 (2006)
6. W K Choi, W K Chim , C L Heng, L W Teo, V Ho, V Ng, D A Antoniadis and E A Fitzgerald , *Appl. Phys. Lett.* **80,** 2014 (2002)
7. Y. Kanemitsu, K. Masuda, M. Yamamoto, K. Kajiyama, T. Kushida, *Journal of Luminescence* **87-89,** 457 (2000)
8. W.K. Choi , S. Kanakaraju , Z.X. Shen, W.S. Li , *Appl. Surf. Sci.* **144–145,** 697 (1999)
9. N.Srinivasa Rao, A P Pathak, G Devaraju and V Saikiran, *Vacuum* **85,** 927 (2011)
10. W K Choi, Y W Ho, S P Ng, V Ng, *J. Appl Phys.* **89,** 2168.(2001)
11. P. Tognini, L.C. Andreani, M. Geddo, A. Stella, P. Cheyssac, R. Kofman, A. Migliori, *Phys. Rev. B* **53,** (11), 6992 (1996) .
12. P. Yadoji, R. Peelamedu, D. Agrawal, R. Roy, *Mat Sci Engg. B* **98**, 269 (2003) .

Mater. Res. Soc. Symp. Proc. Vol. 1354 © 2011 Materials Research Society
DOI: 10.1557/opl.2011.1083

Comparison of ion beam and electron beam induced transport of hot charge carriers in metal-insulator-metal junctions

Johannes Hopster[1], Detlef Diesing[2], Andreas Wucher[1] and Marika Schleberger[1]
[1]Universität Duisburg-Essen, Fakultät für Physik, 47048 Duisburg, Germany
[2]Universität Duisburg-Essen, Fakultät für Physikalische Chemie, 45117 Essen, Germany

ABSTRACT

The generation of hot charge carriers within a solid bombarded by charged particles is investigated using biased thin film metal-insulator-metal (MIM) devices. For slow, highly charged ions approaching a metal surface the main dissipation process is electronic excitation of the substrate, leading to electron emission into the vacuum and internal electron emission across the MIM junction. In order to gain a deeper understanding of the distribution and transport of the excited charge carriers leading to the measured device current, we compare ion induced and electron induced excitation processes in terms of absolute internal emission yields as well as their dependence on the applied bias voltage.

INTRODUCTION

The interaction of charged particles with surfaces has been intensively investigated in the last years. Therefore the main stages of the interaction process are well understood [1-6]. The ions carry energy, which may be both kinetic energy due to their velocity and potential energy due to their charge state. During the impact, ions introduce energy into the surface either via nuclear or electronic stopping. We are interested in the interaction of *slow* highly charged ions (HCI) with metal surfaces, where the main dissipation process is electronic excitation due to elementary Auger processes [7,8]. This excitation, localized in the near-surface region, produces electron emission into the vacuum and internal electron emission in the metal film. The external electron emission into the vacuum is well understood and has been discussed and reviewed in many works before [6,9].

The internal electron emission is based on the transport of excited charge carriers into the bulk of the substrate across an internal energy barrier located below the surface. Since the internal barrier is lower than the work function of the irradiated metal, it is possible to gain new insights into the lower energy part of the excited electron distribution. In this work, we use thin film metal-insulator-metal (MIM) tunnel junctions as sensitive detectors to investigate the electronic excitation created in the top metal film [10]. In order to gain a deeper understanding of the distribution and transport of the excited charge carriers leading to the measured internal emission yields, we compare the device current measured under impact of highly charged ions with that produced during monoenergetic electron bombardment of the surface. By varying the bias voltage between the top and bottom metal electrodes of the MIM device, it is possible to modify the barrier characteristics of the MIM junction. From the resulting bias voltage dependence of the measured device current, we will draw conclusions regarding the ion-induced excitation distribution in the top metal film and the potential energy dependence as well.

The decisive difference between an ion induced excitation and an electron beam induced excitation is the low mass ratio between projectiles and target electrons for the electron induced excitation. Even for target electrons with higher effective masses induced by the band structure of the metal significant portions of momentum and energy can be exchanged in a simple classical scattering model. This is not the case for ion induced excitations with low kinetic energies as presented in our earlier works [10,11,12]. Hence, one can expect decisively different results from ion and electron induced excitations. This motivates our study, where both excitation methods are compared.

EXPERIMENT

In this experiment we use a UHV setup (for details see [13]) equipped with a low energy electron gun for electron irradiation. It can be operated in pulsed mode and the electron energy can be varied from 10 eV up to 1 keV.

The beam - ion or electron - is focused onto a MIM target under normal incidence. The top metal film is about 30 nm thick, the oxide film 3-5 nm and the bottom metal film about 70 nm. The MIM structures are formed by anodic oxidation and may consist of different materials like Al/AlOx/Ag or Ta/TaOx/Au [14]. The primary beam current I_0 is measured with a Faraday cup. For measuring the current through the MIM junction and for applying voltages between top and bottom electrode we use a potentiostat in connection with a current-voltage-converter.

The operation principle of the MIM and its advantage are as follows. The irradiation of the top metal film generates electronic excitation in both cases – ion and electron beam irradiation. The hot charge carriers propagate within the metal film and may eventually reach an interface, metal/vacuum or metal/oxide, depending on wave vector k and energy E. Along its path each electron undergoes inelastic (e-e scattering) and elastic (e-phonon and e-atom scattering) scattering processes characterized by the respective mean free path. This leads to energy loss and/or direction changes and production of secondary electrons, thereby generating a broad distribution of excited charge carriers (electrons and holes) within the top metal film.

When an electron reaches the surface, electron emission into the vacuum ("external emission") is possible if its excitation energy is larger than the work function W of the top metal film. These electrons are easy to detect, but the external electron emission process involves only those electrons excited with sufficient energy ($\approx$ 4-5 eV).

Those electrons, who reach the metal/oxide interface, may have lower energy and can still be detected due to the lower internal barrier φ. Because the MIM junctions used here usually consist of oxide films with about 3-5 nm thickness, the probability for quantum mechanical tunneling through the barrier is negligibly small. Thus, the only way for electrons to reach the bottom metal is to overcome the potential barrier φ_e into the oxide conduction band with typical heights of 1-2 eV for Ta/TaOx/Au devices. In the same way hot holes can reach the bottom metal film by passing "below" the hole barrier φ_h into the oxide valence band. Hence, both kinds of charge carriers can reach the bottom metal film and can be detected there. Note that due to their different polarity the measured device current is a net current $I_m = |I_e| - |I_h|$, which may be positive or negative depending on which channel prevails [10,11].

The exact shape of the potential barrier in a MIM structure depends on the interfaces of the metal films to the interstitial oxide film. In a simplified approach, the Fermi energies of the films align and the barrier heights φ_i are determined by the work function W_i of the metal film and the electron affinity ψ of the oxide, i.e. $\varphi_i = W_i - \psi$ (i: Au, Ta). However, a detailed analysis shows that this picture is largely misleading, with the actual values being essentially determined by the dipole layers at the oxide-metal interfaces. This effect accounts for a difference of 0.7 eV for the MIMs used in this work [15].

If the same metal for top and bottom electrode is used, the simplified model (neglecting the dipole layers) predicts a rectangular barrier with equal barrier heights at both sides. In case of different metal films (like in this work) the barrier heights are different, leading to a potential energy variation depicted schematically in Figure 1. More specifically, an intrinsic *built-in field* is created in the oxide, and as a consequence the barrier is asymmetric and trapezoidal.

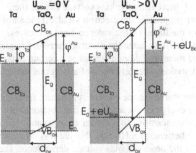

Figure 1 Potential scheme of a Ta/TaO$_x$/Au – MIM for U_{bias} = 0 V (left) and U_{bias} > 0 V (right).

In order to gain information about the energy distribution of excited charge carriers in the top metal we measure the current I_m into the bottom substrate electrode as a function of the bias voltage U_{bias}. When a positive bias voltage U_{bias} > 0 is applied, the position of the top metal Fermi level is raised, and for U_{bias} < 0 the position of the top metal Fermi level is lowered with respect to the bottom Fermi level (see Figure 1).

The beam induced current has to be normalized with respect to the primary beam current I_0 to get the internal emission yield γ. In case of electron beams the yield is $\gamma_{el}=I_m/I_0$, which is the net number of electrons flowing from the top metal to the bottom metal per incident electron. In case of ion beams it is $\gamma_{ion}=(I_m/I_0)\cdot q$, where q is the ion charge state [10,11,12]. If the current I_m is measured with a bias voltage $U_{bias} \neq 0$ a bias induced intrinsic tunneling current has to be subtracted. This current contribution depends on the barrier shape and U_{bias} and can be measured by pulsing the beam at the same bias voltage and measuring the current I_m.

RESULTS AND DISCUSSION

The data of the ion beam induced measurements presented in this paper for comparison has been obtained in previous works [12]. Here we want to show and briefly discuss only the essential points. The yield γ_{ion} provides information about the excited electron distribution. As one can see from Figure 2 (left), the internal emission yield γ_{ion} depends linearly on the potential energy, which is in good agreement with measurements of the external electron emission γ_{ext} due to impact of multiply charged ions on metal substrates [8,9]. Since the deexcitation of impinging ions takes place mainly due to cascading Auger transitions, this process leads to an isotropic electron emission with a broad electron energy distribution.

To learn more about the resulting electron distribution, γ_{ion} was measured as function of the bias voltage U_{bias}. Figure 2 (right) shows data obtained with the same MIM junction for two different charge states, i. e. potential energies of the impinging ions. For a better comparison the data points for both charge states were normalized to the yield at U_{bias}=0 V. It is apparent that a positive voltage increases the yield for both charge states. At first sight, this finding is surprising since the actual barrier height is not influenced by the bias voltage, leaving the number of electrons surpassing the barrier practically unchanged. However, the effective *width* of the barrier changes and becomes smaller with increasing U_{bias}. Therefore, the contribution of electrons tunneling through the barrier with energies slightly below the threshold increases, leading to the observed yield increase. At negative bias voltages, the effect is reversed and the effective barrier width increases, until at -0.7 V the barrier becomes flat and its width corresponds to the entire oxide film thickness [15]. Increasing the bias voltage further then effectively increases the barrier height and further suppresses the electron yield. At the same time, the effective barrier width for hole transport mediated by the valence band decreases, thereby increasing the hole current which acts to counterbalance the electron current. Frequently, this process even leads to a polarity change of the measured device current at a certain bias voltage (as observed in Figure 2 (right) for Ar^{6+}), making the definition of an internal emission yield practically meaningless.

The influence of the bias voltage on the measured device current is different for different charge states. Apparently, ions with low potential energy evoke an energy distribution with predominantly low energy electrons (which are more sensitive to a barrier change) and ions with higher potential energy produce electrons with higher energies ($E > \varphi$), that are less influenced by a barrier change. A more detailed discussion of the observed bias voltage dependence, the reader is referred to our previous publications [11,12].

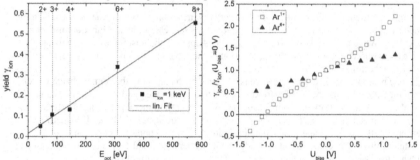

Figure 2 Yield γ_{ion} measurements from Peters et al. [12]. **Left:** γ_{ion} as function of the potential energy E_{pot} for ions with $E_{kin}=1$ keV. **Right:** γ_{ion} as function of the bias voltage U_{bias} for two charge states, normalized to γ_{ion} ($U_{bias}=0$ V).

In order to examine the proposition that ion irradiation leads to electronic excitations with different energies depending on the charge state, we performed experiments with pulsed electron beam irradiations at different impact energies for comparison.

Figure 3a) shows the electron induced internal emission yield γ_{el} for an energy range of 10 eV $\leq E_{kin} \leq$ 1 keV. In the lower energy range below, say, 300 eV, the yield distribution resembles that of secondary electrons measured for external emission into the vacuum. In particular the yield maximum at several hundred eV is generally observed in secondary electron emission as well, although it occurs at higher impact energy (~ 500 eV) for Au [16]. The maxima may reflect structures in the stopping power related to the shell structure of Au [17]. For energies exceeding 600 eV the internal yield measured here strongly deviates from the external emission yield. While the latter continues to decrease, we find a strong increase of the internal yield with increasing impact energy. At about 700 eV, the yield reaches a value of unity, indicating that every impinging electron overcomes the internal barrier and reaches the substrate electrode. At still larger impact energy the yield even exceeds unity.

It is interesting to speculate about the reason for this difference between external and internal emission yields. External emission requires the generation of secondary electrons with momentum directed towards the surface within a sub-surface depth interval essentially determined by the average electron-electron mean free path (0.5 nm for Au [16]). On the other hand, it is well known that the stopping power for the primary electron decreases with increasing impact energy [18], leading to the observed decay of the external emission yield towards high impact energies. Internal emission, on the other hand, is favored by a decreasing stopping power, since it increases the probability for the primary electron to reach and overcome the internal barrier. Ultimately, with the stopping power going to zero, one would therefore expect an internal emission yield of unity in the limit of very high impact energy. The fact that we observe even higher yields must result from inelastic electron-electron scattering processes, which produce secondary electrons with sufficient energy to overcome the barrier as well. It is clear that these processes will be more efficient in the forward scattering direction, favoring internal emission again.

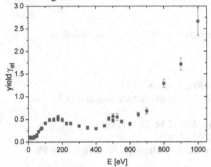

Figure 3 a) Yield γ_{el} as function of the electron beam energy E with $U_{bias}= 0$ V.

Figure 3 b) Yield γ_{el} as function of U_{bias} for different electron beam energies E.

To obtain more information about the energy distribution at the metal/oxide interface we analyzed the yield γ_{el} as a function of the bias voltage U_{bias} for several beam energies (see Figure 3b)). The bias voltage varies from U_{bias}=-0.8 V to U_{bias}=0.6 V. Higher bias values were avoided to prevent the dielectric breakdown of the oxide layer. Particularly interesting here is the linear dependence of the curves, which has been observed for ion induced internal emission as well.

An essential point in Figure 3b) is the difference of the slope for different beam energies, which appears to scale with impact energy. Apparently, the electron energy distribution generated by high-energy electrons is more sensitive to bias-induced barrier changes than that produced by low energy electrons and ions. At low impact energy, one might speculate that the measured yield basically reflects the probability of the primary electron to reach the oxide interface with enough energy to overcome the internal barrier. At higher energy, secondary electrons produced by inelastic energy loss of the primaries start to contribute to the measured yield. The energy distribution of these secondaries must be dominated by low energy electrons, which are more strongly influenced by relatively small changes of the barrier.

CONCLUSION

Our first results with electron beams show a decreasing bias dependence for a decreasing particle energy. This is quite in contrast to the results obtained with ion induced electronic excitation by potential energy [11,12], by kinetic energy [19,20,21] and photo induced electronic excitations as well [15]. All of these excitation methods lead to a decreasing bias dependence for increasing particle energies. One should, however, note the different energy scales involved in those experiments. While the excitation induced by photons or kinetic ion impact must necessarily be dominated by low energy electrons (and holes), an electron impact with hundreds of eV involves energies in a completely different regime. As outlined above, an internal emission yield of unity can be obtained merely by penetration of the primary electron into the substrate. Experiments with silver as top electrode are under way. By exchanging the top metal we will be able to discriminate element specific electronic excitations. The role of scattering and production of secondary electrons will be addressed by a variation of the top metal thickness. For a deeper understanding, a computer simulation of the experimental results is necessary, which includes transport of charge carriers in the top metal and secondary electron excitation in particular.

ACKNOWLEDGMENTS

Financial support by the DFG – SFB616: *Energy Dissipation at Surfaces* is acknowledged.

REFERENCES

1.   J. P. Briand, L. de Billy, P. Charles et al. *Phys. Rev. Lett.* **65**, 159 (1990)
2.   F. W. Meyer, S. H. Overbury, C. C. Havener, P. A. Z. van Emmichoven and D. M. Zehner, *Phys. Rev. Lett.* **67**, 723 (1991)
3.   J. Burgdörfer, P. Lerner and F. W. Meyer, *Phys. Rev. A* **44**, 5674 (1991)
4.   J. Das and R. Morgenstern, *Phys. Rev. A* **47**, R755 (1993)
5.   D. Kost, S. Facsko, W. Möller, H. Hellhammer and N. Stolterfoth, *Phys. Rev. Lett.* **98**, 225503 (2007)
6.   A. Arnau, F. Aumayr, P. M. Echenique et al. *Surf. Sci. Rep.* **27**, 113 (1997)
7.   F. Aumayr, A. S. El-Said and W. Meissl, *Nucl. Instrum. Methods B* **266**, 2729 (2008)
8.   J. D. Gillaspy, *J. Phys. B: At. Mol. Opt. Phys.* **34**, R93 (2001)
9.   H. Winter, *Europhys. Lett.* **18**, 207 (1992)
10.   D. Kovacs, A. Golczewski, G. Kowarik, F. Aumayr and D. Diesing, *Phys. Rev. B* **81**, 075411 (2010)
11.   D. A. Kovacs, T. Peters, C. Haake, M. Schleberger, A. Wucher, A. Golczewski, F. Aumayr and D. Diesing, *Phys. Rev. B* **77**, 245432 (2008)
12.   T. Peters, C. Haake, D. Diesing, D. A. Kovacs, A. Golczewski, G. Kowarik, F. Aumayr, A. Wucher and M. Schleberger, *New Journal of Physics* **10**, 073019 (2008)
13.   T. Peters, C. Haake, J. Hopster, V. Sokolovsky, A. Wucher and M. Schleberger, *Nucl. Instr. Meth. B* **267**, 4 (2009)
14.   D. Diesing, G. Kritzler, M. Stermann, D. Nolting and A. Otto, *J. Solid State Electrochem.* **7**, 389 (2003)
15.   P. Thissen, B. Schindler, D. Diesing and E. Hasselbrink, *New Journ. of Physics* **12** (2010)
16.   Y. Lin and D. C. Joy, *Surf. Interf. Anal.* **37**, 895-900 (2005)
17.   W. Lotz, *Journal of the Opical Society of America* **60**, 2 (1970)
18.   H. Bethe and W. Heitler, *Proceedings of the Royal Society of London* **146**, A856 (1934)
19.   S. Meyer, D. Diesing, and A. Wucher, *Phys. Rev. Lett.* **93**, 137601 (2004)
20.   D. Diesing, D. Kovacs, K. Stella and C. Heuser, *Nucl. Instr. Meth. B*, accepted (in press)
21.   S. Meyer, C. Heuser, D. Diesing and A. Wucher, *Phys. Rev. B* **78**, 035428 (2008)

Forum Report

Mater. Res. Soc. Symp. Proc. Vol. 1354 © 2011 Materials Research Society
DOI: 10.1557/opl.2011.1457

Future Directions for Ion Beam Technology and Research: Forum Report

John E.E. Baglin[1] and Daryush Ila[2]

[1]IBM Almaden Research Center, 650 Harry Rd., San Jose, CA 95120, USA;
[2]University of North Carolina - Fayetteville State University, 1200 Murchison Rd., No. 15475, Fayetteville, NC 28301-4297, USA.

ABSTRACT

As an integral part of the Symposium on "Ion Beams - Applications from Nanoscale to Mesoscale" at the MRS Spring 2011 Meeting, participants were invited to join two open "brainstorming" Forum Discussions, intended to highlight opportunities for application of ion beam techniques in advancing the frontiers of materials research and making high impact contributions to solving some of the world's major issues for the future. Participants were invited to imagine freely how the field might develop (or be steered) in the next 5-10 years, in the light of the current state of the art, and in the light of the emerging needs of the global community.

The resulting ideas and suggestions led to thoughtful discussions, that displayed a remarkable degree of consensus on future directions, opportunities and challenges for the field. This paper attempts to capture and report briefly the spectrum of ideas and the recommended priorities that emerged from the resulting discussions.

OUTLINE

The variety and innovative enterprise displayed in many recent publications and conferences on ion beam technology, science, and applications, indicate a field in which new ideas and insights are stimulating studies in many rapidly evolving topics such as bio materials or nanoscale devices, with the potential for having major impacts in addressing some of the world's big problems and challenges.   It therefore seemed to be an appropriate time for an informal assembly of frontier researchers to exchange their varied ideas and perceptions and thoughtful predictions about the future progress of the field, topics of high potential, and exciting and timely opportunities for research ranging from fundamental discoveries to commercial applications and novel instrumentation. Thus, two special open Forum events were included in the program of the Symposium on "Ion Beams - Applications from Nanoscale to Mesoscale" at the MRS Spring 2011 Meeting. Following the style of similar discussion sessions of recent invitational "Ion Beam Institutes" hosted by Alabama A&M University, participants were invited to briefly express their personal ideas about the progress and future of the entire field and its applications, challenges, and opportunities. J. Baglin served as Moderator.

The following report attempts to describe the context of the discussions, and summarize (without any editorial selection) the great variety of topics and suggestions that arose, reflecting an encouraging consensus of opinion in many cases, among the approximately 50 attendees.

Many of the responses proposed broad areas in which ion beam techniques should be expected to make key enabling contributions. However, some of the responses (appropriately) centered on visions of dramatic advances in instrumentation, software and engineering that now seem likely to be attainable, enabling new ubiquitous applications of ion beam technology. Further broad suggestions covered issues of local and international networking that can facilitate such advances.

A *Forum on High Impact Applications (i.e. "Big-Picture")* was held following an invitational Dinner for Invited Speakers. An *Open Discussion Networking Forum* was the concluding event of the Symposium. Both events were marked by enthusiasm, insight and thinking "outside the box", and both portrayed a field having deep roots in history, that now sees a world of great opportunity for innovation. The two forums were not structured with the intent of generating a coordinated strategic roadmap for the field, or to rank priorities in "wish-lists". Rather, all participants were urged to exercise creative and imaginative thinking, and to freely articulate ideas and visions for possible actions, investigations, or initiatives whereby applications of ion beam based knowledge and technology might develop to have major impact on the future progress of Science and the World.

The suggestions and ideas for fruitful action (compiled from both discussion events) are listed in the concluding Section titled "Future Ideas, Opportunities, Wish Lists and Recommended Actions".

The two events displayed several areas of remarkable consensus on topics considered to be most important, in addition to yielding a stimulating cluster of challenges, and ideas identified for future serious consideration. These abbreviated opinions and proposals are recommended for thoughtful evaluation as research programs and business plans emerge for advanced technologies in the coming decade. An edited summary of the principal recurring themes, ideas and suggestions emerging during these discussions is presented below. This is followed by a more comprehensive list of very concisely stated ideas and proposals for actions that might fruitfully be taken in the near term to support and stimulate progress in various emerging topics.

SUMMARY OF THEMES AND CONSENSUS TOPICS

Opinions, ideas and suggestions for future major growth arose mainly in three categories as grouped in the lists below. We display here (without elaboration) the issues of principal interest and consensus.

1. Biomedical and bioscience and human health applications.

Ion beam techniques can provide critical support for both research and application in fields such as:
- Targeted drug delivery
- Sensors
- Advanced, highly specific, diagnostics
- Molecular mass spectrometry
- Biocompatible materials
- Nanopore fabrication for DNA sequencing

- Study and tailoring of cell surfaces and coatings
- Affordable and ubiquitous therapies

Factors for future success will include:
- Effective collaborations (at levels ranging from fundamental research to adopted implementation) between the bio- and ion-technology communities
- Greater availability of ion beam facilities built to serve these special purposes
- Inexpensive, turn-key tools for research and clinical use.

2. Toward ubiquitous implementation of ion beam technologies.

Continuing efforts are recommended to design, build and market versatile but purpose-designed ion beam accelerators, sources and delivery facilities that are user-friendly, and compact.

– Examine major emerging technology fields (such as renewable energy or photovoltaics) for potentially fruitful introduction of ion beam technologies

– Facilitate research collaborations by users with equipment manufacturers to create new, commercially attractive facilities and software.

– Updated instrumentation and techniques for characterization and metrology, introducing high sensitivity, reliability, chemical imaging, rapid analysis, and suitability for polymers and biological materials.

– Versatile tools for nanoscale prototyping, including multiple beams, cluster beams, ALD facilities, programmable pattern generators, adaptation of projection beam concept, and 3D nanoscale fabrication.

3. Interdisciplinary opportunities.

The need exists for professionally oriented programs in which researchers and practitioners in various fields of specialization may meet with and inform their counterparts with ion beam experience and expertise. A possible example would be a series of Summer Institutes, focused on interfacing among engineering and science fields, e.g., Bio + Nano; Ions + Lithography.

ISSUES, RECOMMENDATIONS, AND WISH LIST ITEMS

In this section, we group the responses to questions such as "What would you recommend for action in the near term that would strengthen the ion beam based research field, and enable its healthiest progress, development and innovation?" Or "What would you wish for, in order to better support the research that you are doing, or would like to do?" These recommendations are grouped in broad topical categories, whose titles are self explanatory.

Enhanced Interactions Proposed

- Initiate more effective interactions with other communities, in such areas as -
    - Characterization of nano-materials
    - Environmental studies
    - Bio medical and biosciences (e.g. cell adhesion, stents, targeted drug delivery, prostheses)
    - Toxicology
    - Radiology
    - Forensics
    - Geosciences
    - Investment and Intellectual Property
    - Emerging industries
- Launch a Summer Institute series, designed to bridge gaps of terminology and science between the ion beam community and other professional groups in engineering, bio technology, government, law, etc.
- Explore new possibilities for ion beam based science and technology to make enabling contributions in such evolving multi-disciplinary fields as renewable energy, photovoltaics, biosciences, etc.
- The need exists to train a next generation of designers and engineers who will take over manufacture of particle accelerators when the current generation retires.
- Develop strong programs to engage student interest, enterprise and research involvement in the newly emerging areas of ion beam science and technology, including real opportunities for students to share new ideas, and to pursue successful career paths.

Actions Suggested

- Develop, publish, and maintain a list of potential funding sources to support technology transfer for emerging ideas, including government agencies, foundations, industries, and venture capital sources.
- List accessible Centers or Programs offering available facilities for visitors / collaborators (e.g. EMSEL, NIMS (Tsukuba, Japan), TNO (Delft, The Netherlands))
- List accessible sources of software, e.g. Kalypso for MD simulation, 3D SRIM, ion beam analysis packages.

Great Ambitions

- Biomedical applications
    - Drug delivery
    - Sensors

- Diagnostics
- Molecular mass spectrometry
- Boron neutron therapy
- Tumor therapy
- Affordable ubiquitous therapy
- Lab-on-chip applications - Bio MEMS
- Ion beam tailoring of organic materials (cells, plant seeds, etc.)

- Fabrication of engineered patterns in arbitrary materials, whose structures, composition, and surface functionalization may be used to influence cell growth or cell attachment (e.g. stem cells, targeted organ growth, or skin).

- DNA sequencing: Ability to create precisely structured nano-pores in thin films by direct ion irradiation with nano-scale beams

- Fabrication of 'smart' nanoporous membranes

- Development of all-photonics devices and systems on a chip.

- Materials Discovery -- Ion beam techniques are ideally positioned to design ways to fabricate speculative materials structures to order.

- Quantum computing; qubit patterning.
    - Deterministic single ion implantation is the best way. Develop tools for this purpose, e.g., sense ions individually in beam line en route to target.

- Maskless, patterned implantation of nanocrystal arrays
    - for photonic / quantum dot applications
    - for hybrid electronic / photonic devices

- Chemical imaging at nanometer scale

- Water purification / desalination via nanopore membranes

- Integrated tools built to address the performance specified in the Semiconductor Industry Roadmap.

- Large area lithography at few-nm resolution

- Routine manufacturing of 3D devices, with high precision, high throughput, and high reliability

- Economically attractive friction/wear/corrosion reduction (on a large industrial scale).

Timely Research Opportunities / Challenges

- Explore influence of ion-patterned nano-dimensional features on mechanical performance (friction, hardness, wear) of surfaces and coatings.

- Cluster ion bombardment effects at the nanoscale - intrinsic kinetics, and surface modification.

- Graphene patterning and lithography, based on nanoscale beam writing.

- Biomimetic materials design (via specific peptides), for tissue engineering / regenerative medicine
- Bio-compatibility for implanted devices
- Fabrication of nanofluidic devices by ion sculpting or lithography
- Microfluidic devices - flow control for lab-on-chip
- Optimize solar-thermal receptor surfaces, properties, and robust structural material.
- Advanced design and fabrication of nanoscale catalysts? Ion patterned growth of nano-features.
- Self-assembled ripples at reduced wavelength and faster reproducible growth rates
- Basic Interactions: Investigate ion-solid interactions over the larger spatial and temporal ranges afforded by systems like the MionLINE.
- Polymer modification and processing:
    - Combined membrane technologies involving ion patterned surface functionalization
    - Ionic-type conducting polymers (actuators for MEMS)
- High density pre-patterned magnetic storage media.
- Defect engineering to generate nanostructures
- Projection ion beam tools. (Make the CHARPAN concept economically viable).
- Plasmonics patterning?
- Nuclear / plasma diagnostics for fusion systems
- Fission/fusion reactor materials. Dedicated multiple-techniques facility? New materials and characterization.
- Microbeam writing using ion species to suit the need, other than Ga.
- Join materials that don't bond naturally - as a routine process..
- Advanced ion propulsion?

Analysis

- Demonstrate simplicity, sensitivity, reliability and precision of well engineered ion beam analysis
- Continue to offer high performance packaged analytical tools for routine industrial application (e.g. AMS for pharmaceutical industry)
- Portable PIXE; Nano-beam PIXE
- Super-sensitive new characterization instruments and facilities; intuitive, quantitative ion beam analysis tools for NRA, ToF-SIMS, D-SIMS, RBS, PIXE, etc., designed to serve many professions (e.g. medical research, industrial quality control, forensic, legal, security areas)

Instrumentation, Resources and Facilities

- Extend ion beam interaction software to include 2D and 3D materials
- More table-top ion beam facilities (not unreasonable with current technology?)
- Partnerships between manufacturers of ion beam equipment and their future customers, to develop new facilities for specific new applications
- Versatile nano-scale prototyping multi-beam tools
- Note a planned multi-tool by Varian, purpose-built for high throughput photovoltaic solar cell wafer production.
- Miniaturized accelerators / ion sources
  - Accelerators designed for non-vacuum sample analysis
  - Table-top single-function tools
  - Affordable implantation (energies up to several MeV)
- Accessible versatile ion beam facilities for broad range of applications (e.g. intense miniature neutron source for radiology or analysis).
- RBS/PIXE capability added to Helium Ion Microscope; higher beam current; alternative ion species.
- Helium-Ion-Microscope with ERDA, for nanoparticle analysis ?
- Convert MionLINE into a deceleration system that writes/deposits nanometer lines for device interconnections. Multi-ion species capability for this tool (as at University of Florida).
- Cluster ion beams qualified to replace CMP in chip processing?
- Single-ion implant systems for quantum computer fabrication
- Z-contrast ion induced microscopy: Select ion, SE yields, channeling angles and exposures to image with high contrast and in real time. Use *in situ* sputter etching to enable depth analysis and imaging with Helium Ion Microscope.

CONCLUDING REMARKS

The ideas, concepts, recommendations and ambitions contained in this summary of the forums serve to illustrate several important features of the current status of Ion Beams research, technology and applications. They include the following:
- The field is vital, dynamic and enterprising - as vividly demonstrated by the selected Proceedings papers published from the MRS Spring 2011 Symposium.
- Development of new instruments and new basic insights during the past decade has stimulated a recognition that ion beam science and techniques now have much more to offer than ever before, especially in the evolving life sciences and in nanoscale engineering for the device industry.
- Strong programs and tangible opportunities are needed at this time, to support and encourage the best and brightest students to consider pursuing careers among the many

diverse, exciting, and dynamic fields of ion beam based science and engineering. The new generation will thus be ready and inspired to value and pursue the wealth of forward-looking recommendations and collective vision such as those presented in this Report.

- One key to exercising the most exciting opportunities and applications is going to lie in actions planned to revolutionize interdisciplinary communications, collaborations, and cross fertilization of inventive ideas and insights among experts in many fields.
- The emergence of purpose-dedicated facilities is already welcomed by industrial customers and collaborators.

One can hope that the wish lists and personal concepts, ideas, and recommendations arising from these forums may serve to inspire further fresh, ingenious and productive programs and techniques in the coming years.

ACKNOWLEDGEMENTS

The authors acknowledge with thanks the many individual participants in these Forum discussions, whose enthusiastically contributed ideas and perspectives constitute the essence of the current Report. The discussions formed an integral part of the Ion Beams Symposium of the Spring 2011 MRS Meeting, which was made possible by generous sponsorship from National Electrostatics Corp., and from NASA, through its contract with the Center for Irradiation of Materials, Alabama A&M University. We also wish to acknowledge the important role of our Symposium Co-Organizers, Giovanni Marletta (Università degli Studi di Catania, Italy), and Ahmet Öztarhan, (Ege University, Izmir, Turkey).

AUTHOR INDEX

SUBJECT INDEX

Printed in the United States
by Baker & Taylor Publisher Services

Printed in the United States
by Baker & Taylor Publisher Services